LAURA CANDLER'S

MASTERING MATH FACTS

MULTIPLICATION & DIVISION
ALIGNED WITH THE COMMON CORE

LAURA CANDLER'S
MASTERING MATH FACTS

MULTIPLICATION & DIVISION
ALIGNED WITH THE COMMON CORE

COMPASS

A DIVISION OF BRIGANTINE MEDIA

Published by Compass, an imprint of Brigantine Media
211 North Avenue, Saint Johnsbury, Vermont 05819

Cover and book design by Jacob L. Grant

Printed in Canada

To order multiple copies of the physical book or the digital download, or for more
information about Compass educational materials, please contact:

Compass Publishing
211 North Avenue
Saint Johnsbury, Vermont 05819
Phone: 802-751-8802
E-mail: neil@brigantinemedia.com
Website: **www.compasspublishing.org**

ISBN: 978-1-9384061-9-5

Acknowledgments

I'm excited to be able to share this revised and updated version of *Mastering Math Facts* with you! It's an accomplishment that I could not have achieved on my own, and I want to express my sincere appreciation to those who assisted me with helping my vision become a reality. I'm grateful to the field testers who worked with me on the original manuscript several years ago, as well as those who contributed to this new version. Teachers who used the program and gave me feedback are listed below.

I want to thank my friend and colleague, Pat Calfee, who collaborated with me on Chapter 1. Pat is an outstanding educator, with expertise in developing lessons that actively engage students while teaching for deep understanding of essential concepts. After I revised Chapter 1, I asked Pat to review it and offer suggestions for additional teaching strategies. She graciously spent several hours helping me tweak that section to make it easy to introduce multiplication and division concepts in a step-by-step manner. She also shared two of her own favorite lessons with me, "Cookies and Chips" and "Hula-Hoop Multiplication," and I've included those activities with her permission.

I would also like to thank my daughter, Wendy Candler, for her artwork in Chapter 2. She created all the ice cream and sundae patterns and did a fabulous job!

Finally, I want to express my appreciation for the hard-working, dedicated team at Compass Publishing. I appreciate Neil Raphel for mapping out the scope of the project and handling the logistics of every aspect of its implementation. I'm grateful to my editor, Janis Raye, who has once again done a superb job of revising and editing my writing so that's clear and easy to understand. Thanks also to Jacob Grant, the graphic designer who brought these pages to life with an eye for simple yet elegant formatting and a talent for creating beautiful illustrations. The whole team at Compass Publishing is amazing, and I'm grateful to them for their encouragement and support over the last year.

Laura Candler

Amy Bailey
Fort Wayne, Indiana

Suzy Brooks
Falmouth, Massachusetts

Kerry Brown
Norwalk, Connecticut

Donna Casino
Schenectady, New York

Suzann Falgione
Fayetteville, North Carolina

Sue Gamm
Gilroy, California

Jennifer Graff
North Bay, Ontario

Nikki Houchins
Fayetteville, North Carolina

Janice Krahenbuhl
Gilroy, California

Francie Kugelman
Los Angeles, California

Kathy Kuny
Wexford, Pennsylvania

Deborah Lenz
Austin, Texas

Sharon Manka
Plumas Lake, California

Debbie Palmer
Mission Viejo, California

Susan Revill
Frost, Texas

Beth Shmagin
Culver City, California

Elizabeth Supan
Aiken, South Carolina

Lorraine Vasquez
San Antonio, Texas

Donna Wilder
Lancaster, South Carolina

contents

The Need to Master Math Facts

• • • • • • • • • •

Most classroom teachers would agree that knowing basic math skills is a major factor in strong mathematics progress, and research confirms this view.

Students who know the basic math facts are more likely to experience success in higher-level mathematics education. The Final Report of the National Mathematics Advisory Panel (2008, U.S. Department of Education) repeatedly refers to "the mutually reinforcing benefits of conceptual understanding, procedural fluency, and automatic (i.e. quick and effortless) recall of facts."

The Common Core State Standards for Mathematics emphasize the importance of students not only developing a conceptual understanding of multiplication and division, but also memorizing those math facts. The last sentence in Standard 3.OA.C.7 is: "By the end of Grade 3, *know from memory* all products of two one-digit numbers" (italics mine).

These experts agree that it is critical that our students memorize math facts. But why? Most children have easy access to calculators, and many adults don't remember the times tables. In today's world, why is memorizing math facts so necessary? The answer is fluency. Fluency is just as important to math success as it is to reading. Students who struggle to decode each word will never improve their reading comprehension until they improve their reading fluency. In the same way, students who are counting times tables on their fingers are doomed to fail in mathematics. How can they learn to divide or reduce fractions when all their mental powers are devoted to recalling basic math facts? According to the Common Core Standards, fourth graders not only have to use multiplication and division in word problems, but they must be able to find all the factors of any number from 1 to 100. Finding factors of a number is extremely time-consuming if you have to rely on a calculator—or your fingers—to do the work. The fifth grade Common Core Standard 5.NBT.B.5 states that students must be able to "Fluently multiply multi-digit whole numbers using the standard algorithm." Fluency is key.

When students haven't mastered the basics, they begin to struggle with many other math concepts. They lose confidence in themselves, think they are not smart and, worse, think they will never be good in math. One by one, the doors of future math and science opportunities begin to close to these students.

One reason many students never successfully master the times tables is that they lack a solid foundation of basic multiplication concepts. Perhaps they were taught through rote memorization without being given the opportunity to explore multiplication concepts with concrete materials. Or maybe they learned the concepts through hands-on exploration but did not spend enough time practicing these skills to develop fluency. Many children are frustrated and bored by lengthy drill and practice worksheets. These students often give up on learning the math facts, particularly if they don't understand *why* they need to master them.

No matter what grade we teach, we have to "own" the problem of poor computational fluency in our students and *insist* that they learn basic math facts. Of course they should have learned the math facts before coming to our classrooms—but it's our responsibility to make sure they learn them before being promoted to the next grade. If they don't have the foundation in place, we have to provide them with opportunities to explore multiplication with hands-on materials. If they need more opportunities to develop fluency, we must engage them in practice activities and multiplication games on a regular basis. We must convince our students of the importance of knowing math facts so they will see math facts mastery as a worthy accomplishment.

Mastering Math Facts is a proven system you can use to help students develop both a deep understanding of basic facts and fluency with them.

Common Core Alignment

Every aspect of *Mastering Math Facts* is aligned with Common Core Math Standards. Chapter 1 consists of eighteen different lessons, each aligned with one or more third grade Content Standards as well as one or more Mathematical Practice standards. Chapters 2, 3, and 4 are all aligned with Content Standard 3.OA.C.7: Fluently multiply and divide within 100. If you teach fourth or fifth grade, you are responsible for making sure students master all math content up to and including your grade. If your students have not met these third grade standards, those standards should be addressed before you can move on to more advanced math concepts appropriate for your grade level.

Using This Book

Mastering Math Facts: Multiplication and Division is primarily for third through fifth grades. A third grade teacher can expect to spend weeks of class time developing basic concepts and should not expect students to become fluent until the end of the year. But fifth grade students have already received instruction in multiplication and division concepts, so they may need only a quick review of the basic concepts. With just 10 or 15 minutes a day of review and practice activities, they can achieve mastery of the concepts and build fluency.

Review the components of the program to decide where you need to begin:

COMMON CORE CONNECTIONS CHART ● This chart (pages 4 - 5) is a quick reference guide that includes all Common Core Standards for math content and mathematical practices that are aligned with specific lessons and activities in the *Mastering Math Facts* program.

CHAPTER 1 ● TEACHING MULTIPLICATION AND DIVISION FOR UNDERSTANDING

Chapter 1 has detailed Common Core-aligned lessons to introduce students to basic multiplication and division concepts. It's critical that students understand what multiplication and division mean before you begin the practice activities. Third grade teachers can expect to spend a lot of time with these hands-on activities. Fourth and fifth grade teachers may want to use the *Multiplication: Show What You Know* assessment first to determine how much time, if any, they need to spend on these activities.

CHAPTER 2 ● THE MASTERING MATH FACTS SYSTEM

Chapter 2 includes a full explanation of the Mastering Math Facts system, developed and used successfully for more than 10 years to help students achieve fluency with the times tables. If your students already understand multiplication and division concepts, you can begin your instruction with the strategies described in this chapter. These activities are designed to take no more than 10 to 15 minutes a day.

CHAPTER 3 ● MATH FACTS ASSESSMENT STRATEGIES

Chapter 3 explains how to assess your students' fluency, both with individual math facts as well as when all the math facts are mixed together. You'll learn how to administer Daily Quick Quizzes and other math facts tests.

CHAPTER 4 ● MATH FACTS PRACTICE ACTIVITIES

Chapter 4 includes a variety of math games and activities that can be used in conjunction with the Mastering Math Facts system to help students achieve mastery. These activities can also be used throughout the year to help students maintain fluency with the times tables.

SUPPLEMENTARY ONLINE RESOURCES ● Additional printables that add "extras" to the program are available at **www.lauracandler.com/mmf**. All downloadable resources are noted with this icon:

Decide where you need to begin with your students, and get started. Let's join forces and declare war on computational illiteracy!

MATH CONTENT STANDARDS (3RD GRADE)	LESSONS AND ACTIVITIES
Content.3.OA.A.1 Interpret products of whole numbers, e.g., interpret 5 × 7 as the total number of objects in 5 groups of 7 objects each.	Hula-Hoop Multiplication (p. 17) Cookies and Chips (p. 19) Egg Carton Groups (p. 23) Fishbowl Multiplication (p. 26) Do the Math: Multiplication (p. 32) Picture This: Multiplication (p. 34) Object Arrays (p. 41) Linking Cube Arrays (p. 43) Graph Paper Arrays (p. 45) Rectangle Race (p. 48) Multiplication Table Arrays (p. 52)
Content.3.OA.A.2 Interpret whole-number quotients of whole numbers, e.g., interpret 56 ÷ 8 as the number of objects in each share when 56 objects are partitioned equally into 8 shares, or as a number of shares when 56 objects are partitioned into equal shares of 8 objects each.	Fair Shares Exploration (p. 62) Fishbowl Division (p. 64) Division Mix-up (p. 74) Do the Math: Division (p. 82) Picture This: Division (p. 85)
Content.3.OA.A.3 Use multiplication and division within 100 to solve word problems in situations involving equal groups, arrays, and measurement quantities, e.g., by using drawings and equations with a symbol for the unknown number to represent the problem.	Do the Math: Multiplication (p. 32) Picture This: Multiplication (p. 34) Do the Math: Division (p. 82) Picture This: Division (p. 85)
Content.3.OA.A.4 Determine the unknown whole number in a multiplication or division equation relating three whole numbers. For example, determine the unknown number that makes the equation true in each of the equations 8 × ? = 48, 5 = _ ÷ 3, 6 × 6 = ?	Linking Cube Arrays (p. 43) Fishbowl Division (p. 64) Mystery Multiplication (p. 177)
Content.3.OA.B.5 Apply properties of operations as strategies to multiply and divide. Example: If 6 × 4 = 24 is known, then 4 × 6 = 24 is also known. (Commutative property of multiplication.)	Linking Cube Arrays (p. 43) Multiplication Table Arrays (p. 52) Math Fact Families (p. 77)
Content.3.OA.B.6 Understand division as an unknown-factor problem. For example, find 32 ÷ 8 by finding the number that makes 32 when multiplied by 8.	Fishbowl Division (p. 64)
Content.3.OA.C.7 Fluently multiply and divide within 100, using strategies such as the relationship between multiplication and division (e.g., knowing that 8 × 5 = 40, one knows 40 ÷ 5 = 8) or properties of operations. By the end of Grade 3, know from memory all products of two one-digit numbers.	Rectangle Race (p. 48) Introducing Math Facts (p. 56) Math Fact Families (p. 77) Mastering Math Facts System – Chapter 2 (p. 97) All assessments in Chapter 3 (p. 139) All math facts practice activities in Chapter 4 (p. 173)

MATH PRACTICE STANDARDS (ALL GRADES)	LESSONS AND ACTIVITIES
Practice.MP1 Make sense of problems and persevere in solving them.	Do the Math: Multiplication (p. 32) Picture This: Multiplication (p. 34) Do the Math: Division (p. 82) Picture This: Division (p. 85) Mystery Multiplication (p. 177)
Practice.MP4 Model with mathematics.	Hula-Hoop Multiplication (p. 17) Cookies and Chips (p. 19) Egg Carton Groups (p. 23) Fishbowl Multiplication (p. 26) Do the Math: Multiplication (p. 32) Picture This: Multiplication (p. 34) Object Arrays (p. 41) Linking Cube Arrays (p. 43) Graph Paper Arrays (p. 45) Rectangle Race (p. 48) Multiplication Table Arrays (p. 52) Introducing Math Facts (p. 56) Fair Shares Exploration (p. 62) Fishbowl Division (p. 64) Division Mix-up (p. 74) Math Fact Families (p. 77) Do the Math: Division (p. 82) Picture This: Division (p. 85)
Practice.MP5 Use appropriate tools strategically.	Linking Cube Arrays (p. 43) Graph Paper Arrays (p. 45) Multiplication Table Arrays (p. 52) Introducing Math Facts (p. 56) Mystery Multiplication (p. 177)
Practice.MP6 Attend to precision (in mathematical communication).	Hula-Hoop Multiplication (p. 17) Fishbowl Multiplication (p. 26) Object Arrays (p. 41) Linking Cube Arrays (p. 43) Graph Paper Arrays (p. 45) Multiplication Table Arrays (p. 52) Fishbowl Division (p. 64) Division Mix-Up (p. 74) Math Fact Families (p. 77)
Practice.MP7 Look for and make use of structure.	Object Arrays (p. 41) Linking Cube Arrays (p. 43) Graph Paper Arrays (p. 45) Multiplication Table Arrays (p. 52) Introducing Math Facts (p. 56) Math Fact Families (p. 77) Mystery Multiplication (p. 177)

CHAPTER 1

• • • • • • • • • •

Teaching Multiplication
and Division
for Understanding

Teaching Multiplication and Division for Understanding

• • • • • • • • • •

The first step towards math fact mastery is to build a foundation of understanding by introducing multiplication and division concepts slowly, using a variety of hands-on activities. Introducing these concepts in many different ways encourages students to develop flexibility in their mathematical thinking. By the beginning of third grade, most children can grasp the idea of multiplication as a shortcut for describing groups of things. But students continue to need plenty of opportunities to explore multiplication and division concepts with real objects and drawings to develop a solid foundation.

The Common Core State Standards place development of this foundation in third grade. The Operations and Algebraic Thinking Standards for third grade include seven specific content standards for teaching multiplication and division. The eighteen lessons in this chapter are all correlated to at least one of these Content Standards. These lessons are also aligned with one or more Mathematical Practice Standards for K-12. The Common Core Connections chart (pages 4 - 5) shows the lessons that are aligned to each Standard, and the teacher directions for the activities and lessons note the Common Core Standards they address.

Because many students have not received this type of instruction in the early grades, upper elementary teachers may need to teach multiplication and division concepts. These teachers should assess their students' prior knowledge of multiplication concepts and choose appropriate hands-on activities to build that missing foundation. Requiring students to memorize times tables when they don't understand them is a big mistake. Teachers can begin introducing basic division concepts using manipulatives after students understand multiplication on a conceptual level, even if they have not developed fluency with the math facts.

Assessing Prior Knowledge of Multiplication Concepts

If you aren't sure what your students already know about multiplication, have each student complete the pretest *Multiplication: Show What You Know* (page 11). This is a quick assessment designed to show if students understand the connection between addition and multiplication, and whether or not they are able to model these concepts.

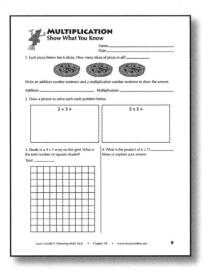

There is a sample student response on page 12, but use your own judgment about accepting answers that are slightly different. For example, when students draw a picture to solve the addition and multiplication number sentences for the second problem, they don't have to draw the same object for both illustrations. However, with the addition sentence, their illustrations should clearly show two items in one set and three more items in another set. When illustrating the multiplication problem they might show any of the following: a 3 x 2 array, a 2 x 3 array, two groups of three objects, or three groups of two objects. Due to the commutative property of multiplication, any of these arrangements would be acceptable. However, if they just drew six objects and didn't show any way of grouping the objects, they are probably missing the concept that multiplication means groups.

Because this is a formative assessment, there's no need to record a number grade on each students' paper. Simply mark their answers correct or incorrect to help you determine who needs more work with the basic multiplication concepts. If you do want to write a number grade on the assignment, you can count each of the four parts as 25% of the test. This might be helpful if you plan to administer the second form of assessment, *Multiplication: Show What You Know 2* (page 13), and want to track their improvement. I do not recommend sending the pretest home to parents or including their percent correct with your other math grades. These grades don't reflect what students have learned in class; they only indicate your students' prior understanding of a concept you have not yet taught.

MULTIPLICATION
Show What You Know

Name:_____

Date:_____

1. Each pizza below has 6 slices. How many slices of pizza in all?_____.

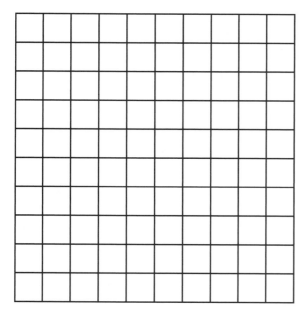

Write an addition number sentence and a multiplication number sentence to show the answer.

Addition:_____   Multiplication:_____

2. Draw a picture to solve each math problem below.

2 + 3 =

2 x 3 =

3. Shade in a 4 x 5 array on this grid. What is the total number of squares shaded?

Total: _____

4. What is the product of 6 x 7?_____ Show or explain your answer.

1. Each pizza below has 6 slices. How many slices of pizza in all? _____18_____.

Write an addition number sentence and a multiplication number sentence to show the answer.

Addition: _____6 + 6 + 6 = 18_____   Multiplication: _____3 X 6 = 18_____

2. Draw a picture to solve each math problem below.

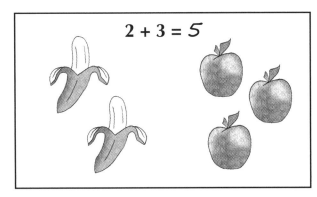

2 + 3 = 5

2 x 3 = 6

3. Shade in a 4 x 5 array on this grid. What is the total number of squares shaded?

Total: _____20_____

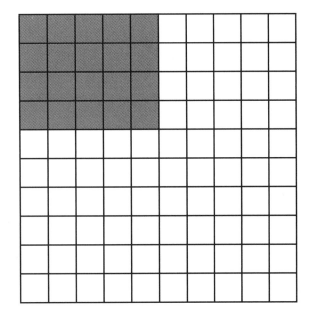

4. What is the product of 6 x 7? _____42_____
Show or explain your answer.

MULTIPLICATION
Show What You Know 2

Name:_____

Date:_____

1. Each lucky clover has 4 leaves. How many leaves in all?_____ .

Write an addition number sentence and a multiplication number sentence to show the answer.

Addition:_____ Multiplication:_____

2. Draw a picture to solve each math problem below.

5 + 3 =	**5 x 3 =**

3. Shade in a 6 x 3 array on this grid. What is the total number of squares shaded?

Total: _____

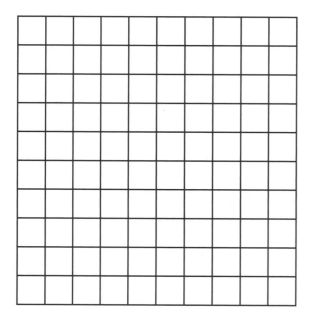

4. What is the product of 4 x 7?_____
Show or explain your answer.

1. Each lucky clover has 4 leaves. How many leaves in all?_____20_____

Write an addition number sentence and a multiplication number sentence to show the answer.

Addition: _4 + 4 + 4 + 4 + 4 = 20_ Multiplication: ___5 X 4 = 20___

2. Draw a picture to solve each math problem below.

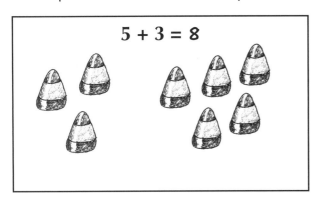

$5 + 3 = 8$

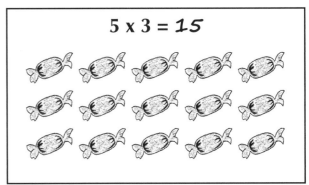

$5 \times 3 = 15$

3. Shade in a 6 x 3 array on this grid. What is the total number of squares shaded?

Total: ____18____

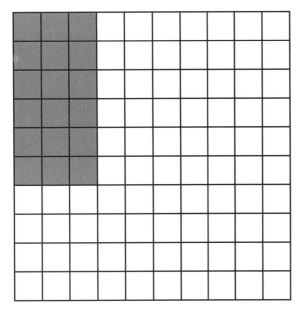

4. What is the product of 4 x 7?_____28_____ Show or explain your answer.

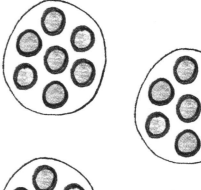
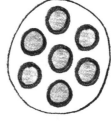
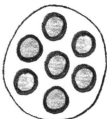

Six Essential Steps to Introduce Multiplication and Division Concepts

Teachers use a variety of techniques for introducing multiplication and division to students, but many of them would agree that there are six essential steps in the process:

1 CONNECT ADDITION AND MULTIPLICATION • Introduce the concept that multiplication is a shortcut for repeated addition and that multiplication involves groups of objects.

2 APPLY MULTIPLICATION TO REAL-WORLD PROBLEMS • Help students make the connection from manipulatives and models to real-life uses of multiplication in word problems.

3 REPRESENT MULTIPLICATION WITH ARRAYS • Explore the ways that multiplication can be symbolized by arrays of objects or squares on a grid, which easily translates to the multiplication table with its grid representation.

4 TEACH ALL MULTIPLICATION FACTS • Teach all of the multiplication facts individually and in a logical sequence. Provide opportunities for students to study and practice them to achieve mastery. Encourage students to memorize multiplication facts and develop fluency with all of the times tables.

5 CONNECT MULTIPLICATION AND DIVISION • Introduce division as the inverse, or opposite, of multiplication. Provide opportunities for students to explore the relationship between multiplication and division using manipulatives and models.

6 APPLY DIVISION TO REAL-WORLD PROBLEMS • Help students apply their knowledge of basic division concepts to real-world problems.

Customizing the Scope and Sequence of Instruction

The lessons and activities that follow are organized according to this six-step process, but you don't have to complete all of the activities in this chapter. Third grade teachers may use almost of all of the strategies, but may vary the order slightly, depending on the needs of their class. Upper elementary teachers who find that their students already understand these concepts should feel free to skip most of the activities in Chapter 1 and move on to the Mastering Math Facts system in Chapter 2.

On the other hand, if you do have some students who were unsuccessful with the pretest questions, I suggest working with them using some of the activities in Chapter 1 while you allow the others to play the multiplication review games in Chapter 4. After re-teaching the basics, you can use the second form of the assessment, *Multiplication: Show What You Know 2* (page 13), to determine when students have mastered the key concepts.

STEP 1 Connect Addition and Multiplication

The most natural way to introduce multiplication is to show students the connection between skip counting in addition and the concept of multiplication as groups of things. These activities will help your students make that transition. The four activities are sequenced in order of difficulty, so be sure to introduce them in this order. You may not need to use all of them—use the pretest results to decide which activities are right for your class.

Each activity is aligned with one or more Common Core Math Content and Practice Standards. The specific Standards covered are noted on the teacher information page for each activity.

- **HULA-HOOP MULTIPLICATION (PAGE 17)**
- **COOKIES AND CHIPS (PAGE 19)**
- **EGG CARTON GROUPS (PAGE 23)**
- **FISHBOWL MULTIPLICATION (PAGE 26)**

Hula-Hoop Multiplication

Common Core Standards:
3.OA.A1
MP4, MP6

After students have mastered skip counting, they are ready to learn about multiplication. This activity works well to introduce multiplication because it involves students physically in an engaging, kinesthetic lesson.

Approximate Time:
20 minutes

Materials ✏️ ✂️ 📎

- At least 6 Hula-Hoops
- Individual dry-erase boards and markers for each student

LAURA'S **Tips**

Dry-erase boards can be made by cutting a large sheet of shower board into 12 x 12-inch pieces. Most home improvement stores will cut the boards for a minimal fee. Small black socks work well as erasers.

Procedure

1 Start by placing 3 Hula-Hoops on the floor (not overlapping) with students gathered around so that everyone is able to see the hoops. Students should be seated with their dry-erase boards, markers, and erasers handy.

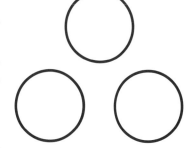

2 Call on 6 volunteers to help with this activity. Ask 2 students to step inside each of the 3 hoops. As you point to each Hula-Hoop, ask the class to skip count aloud with you to find the total. As they count, "2, 4, 6," write the addition number sentence on the board that represents the problem (2 + 2 + 2 = 6). Ask students to copy the number sentence on their dry-erase boards.

3 Repeat this procedure several times with different numbers of hoops and students, but keep the numbers small, never using more than four hoops until the next step. For each round, have your students skip count to find the total and then write the addition number sentence on their dry-erase boards.

4 After several rounds with small numbers of hoops, place six Hula-Hoops on the floor and ask two students to step into each hoop. Ask them to write the addition number sentence for this arrangement. After they write 2 + 2 + 2 + 2 + 2 + 2 = 12, say, "Wow! That's a really long addition sentence! I wonder if there's an easier way to write this number sentence?" Ask if anyone knows another way to show the problem before you reveal the answer; some of the children may already know

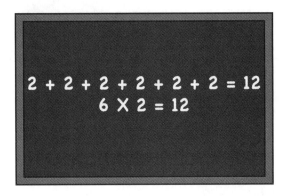

about multiplication and they will be excited to share. If no one knows, tell the students that there's another kind of number sentence they can write that means the same thing. The operation is called *multiplication* instead of *addition*, and it uses an X for the symbol instead of the plus sign. Explain that the first number means the number of groups (Hula-Hoops) and the second number means the number of things in each group (students). Write "6 x 2 = 12" on the board and explain that this is a shortcut for writing the long list of addends in the above problem. Have them copy both the addition and the multiplication number sentences on their dry-erase boards.

5 Repeat step 4 with different numbers of Hula-Hoops and different numbers of students in each hoop. Have students write both the addition number sentence and the multiplication sentence for each round of the activity.

LAURA'S **Tips**

Place only the number of hoops you need on the floor at any one time. It's confusing for young children to see more hoops on the floor than you need for the activity.

Cookies and Chips

Common Core Standards:
3.OA.A.1
MP4

After students have participated in a whole-class introduction like *Hula-Hoop Multiplication*, they are ready to work with partners in a guided math activity. Model this activity in front of the class using a document camera, an interactive whiteboard, or an overhead projector.

Teacher Preparation

Copy, laminate, and cut out enough paper cookies (pattern on page 21) so every pair of students has 6 cookies (3 cookies for every student in your class). The cookies can be cut out exactly or, to save time, along the dotted lines. For the "chips," use centimeter cubes, plastic chips, or dried beans. Place 6 paper cookies, about 40 chips, and 1 die into each plastic bag. Make enough bags so each pair of students has one. Cut out one set of *Cookies and Chips Multiplication Sentences* (page 22) and make a pile of them on your desk.

Approximate Time:
45 minutes

Materials ✎ ✂ 📎

- 1 set of *Cookies and Chips Multiplication Sentences* (page 22)

- 6 paper cookies (page 21) for each pair of students

- 40 chips or small objects for each pair of students

- 1 die for each pair of students

- 1 plastic zip bag for each pair of students

- 1 dry-erase board and marker for each student

Whole-Group Lesson Procedure

1 Pair up your students and give each pair a plastic bag with the *Cookies and Chips* materials (cookie patterns, 40 chips, and a die).

2 Tell students that they are going to learn about multiplication with an activity called *Cookies and Chips*. Assign one person as the Cookie Captain and the other as the Chip Captain. They will switch roles during the activity.

3 Draw one multiplication sentence from your pile and show it to your students. Tell them that the first number is the number of cookies and the second number is the number of chips on each cookie. Their job will be to find out how many chips there are on all the cookies.

4 Using 3 x 2 as an example, ask the Cookie Captains to take 3 cookies out of their bags and place them on their desks. Then ask their partners, the Chip Captains, to place 2 chips on each cookie. Demonstrate using the document camera or overhead projector and transparent cookies.

5 Ask, "How many chips in all?" Have students skip count to find the total, and write two number sentences on their dry-erase boards to represent the total number of chips: 2 + 2 + 2 = 6 and 3 x 2 = 6.

6 When finished, students remove the chips and put the cookies and chips back in the bags.

7 Draw out a new multiplication sentence from your pile and show it to the class. Remind them that the first number means "how many cookies" and the second number is "how many chips." Have them model the solution as they did before and then write both the number sentence and the multiplication sentence on their dry-erase boards. Walk around and assist as needed.

8 After you have completed two or three more rounds of the activity, have students switch roles. The Cookie Captain becomes the Chip Captain and vice versa. Continue to play until students are comfortable with the concepts and procedure. Now it's time to play the *Cookies and Chips* game.

Practicing with *Cookies and Chips* Game

1 Tell students they are going to play the *Cookies and Chips* game. Explain that this is a *learning* game, not a competitive game, and there are no winners and losers. Have them take out the die from their bags and designate one partner to begin. Talk them through the directions the first time they play.

2 Player 1 rolls the die once to find out how many cookies are needed and he or she places that number of cookies between the players. Then Player 2 rolls the die and places that number of chips on each cookie. They both count to find out how many chips in all.

3 On their dry-erase boards, each student writes the addition number sentence and the multiplication number sentence that shows the solution. They compare their dry-erase boards and discuss their answers.

4 Players clear the desk of cookies and chips and play again, switching roles each time they play. Remind them that they are not competing against each other so they don't score points during this activity.

Cookies and Chips Patterns

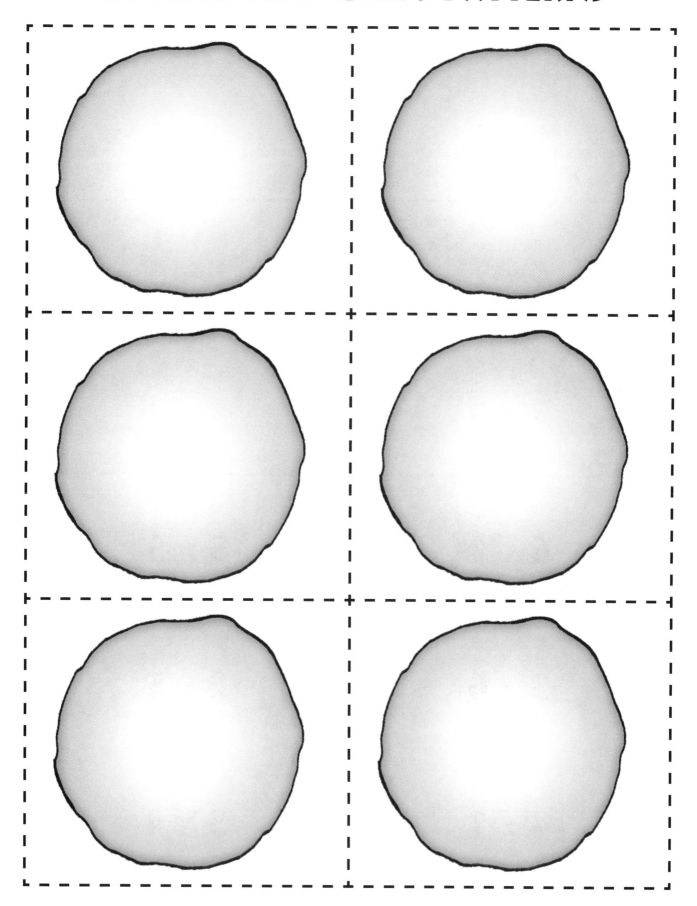

Cookies and Chips
Multiplication Sentences

3 x 5	2 x 6
4 x 2	1 x 5
5 x 4	3 x 2
6 x 4	5 x 5
3 x 6	2 x 5
4 x 1	3 x 4

Egg Carton Groups

Common Core Standards:
3.OA.A.1
MP4

Egg Carton Groups is a little more challenging than *Cookies and Chips* because the egg carton has 12 compartments. The fact that some of the compartments remain empty during the activity can be confusing to young children. You can model the steps of the activity in front of the class using a document camera or by drawing two rows of six circles on an interactive white board to represent the 12 compartments.

Teacher Preparation

Have enough empty egg cartons for half the number of students in your class. Count out 100 small manipulatives, such as plastic chips or beans, for each pair of students in your class. Copy (and laminate) enough Egg Carton Groups cards (page 25) for every pair of students and cut out the individual cards, having 1 set of cards for each pair of students.

Materials

- 1 egg carton for each pair of students
- 100 small manipulatives (e.g., dried beans or plastic chips) for each pair of students
- 1 set of Egg Carton Groups cards (page 25) for each pair of students
- 1 dry-erase board and marker for each pair of students

Approximate Time:
45 minutes

Whole-Group Lesson Procedure

1 Seat students in pairs and give each pair one egg carton, a bag of at least 100 small manipulatives, and one set of Egg Carton Groups cards (page 25).

2 Explain that multiplication is a way of counting groups of things and the egg carton will help them practice their multiplication skills.

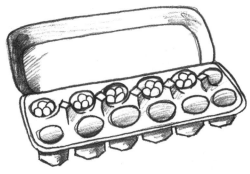

3 Draw one of the Egg Carton Groups cards and read it aloud. Then count out the number of beans to according to the phrase on the card. As each group of beans is added to the egg carton, students skip count and name the total number of beans.

4 For example, if the card says, "5 groups of 3," add 3 beans to each of 5 egg carton cups. As you do this, have students skip count aloud with you: "3, 6, 9, 12, 15."

5 Using the dry-erase boards, each student writes both the addition sentence and the multiplication sentence for the arrangement of objects in the egg carton. In this case, the addition sentence is $3 + 3 + 3 + 3 + 3 = 15$ and the corresponding multiplication sentence is $5 \times 3 = 15$.

Partner Practice

1 After modeling this for the whole group, ask the pairs of students to take turns being the Caller and the Counter.

2 The first Caller draws out an Egg Carton Group card and announces the phrase (for example, "4 groups of 5").

3 The Counter adds the necessary dried beans to the egg carton cups and skip counts to find the total.

4 Both students write the addition sentence and the multiplication sentence on their dry-erase boards, compare their work, and discuss.

5 Students switch roles and repeat the activity as time allows.

Egg Carton Groups Cards

4 groups of 2	6 groups of 3
5 groups of 4	3 groups of 5
9 groups of 2	7 groups of 3
8 groups of 5	6 groups of 2
5 groups of 3	3 groups of 4
2 groups of 7	5 groups of 2

Fishbowl Multiplication

Common Core Standards:
3.OA.A.1
MP4, MP6

Fishbowl Multiplication will reinforce the concept of groups. The "fish" in this activity can be any small manipulative—paper clips, plastic chips, or unit cubes. The game does not require students to write down the number sentences, but you can have students do that in a math journal or on dry-erase boards, if you prefer. Begin the activity by modeling it for the whole class, and then let students practice with a partner or in a math center.

Teacher Preparation

You'll need one bowl for each pair of students. Fill each bowl with about 30 "fish." Copy and laminate 1 *Fishbowl Multiplication* game board (page 29) for each pair of students. Copy, laminate, and cut apart the Addition and Multiplication Sentence cards (page 30), 1 set for each pair of students. Copy 1 set of *Fishbowl Multiplication* directions (page 28) for each pair of students.

Materials ✏️ ✂️ 📎

- 1 *Fishbowl Multiplication* game board (page 29) for each student pair

- 1 set of *Fishbowl Multiplication* game directions (page 28) for each student pair

- Addition and Multiplication Sentence cards (page 30) for each student pair

- Small bowl filled with about 30 "fish" (paper clips, plastic chips, or unit cubes for each student pair

Approximate Time:
45 minutes

Whole-Group Lesson Procedure

1 Place the addition sentences face down in a pile. Spread out the multiplication sentences face up on the table. Start with all the fish in the bowl.

2 Choose two students to help you model the game in front of the class using a document camera or overhead projector. Project the game board. Partner A chooses an addition sentence and places it face up in the box at the bottom of the page. He or she fills the "fishbowls" on the game board with equal groups of "fish" according to the card.

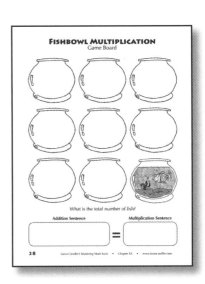

3 Partner B finds the matching multiplication sentence and places it face up in the other box. Partner B then counts to find the total number of fish and announces it to Partner A.

4 If Partner A agrees, they remove both cards and set them aside. Put all the fish back in the bowl. If not, recount the fish.

5 Repeat steps 1 - 4, with Partner B choosing the addition sentence and Partner A finding the multiplication sentence.

6 Repeat this one or two more times, making sure the whole class understands the game.

Partner Practice

1 Pair students and give each pair a game board, a bowl of manipulatives, and a set of game cards. Display the directions (page 28) or give each pair a set of printed directions.

2 Designate who is Partner A and who is Partner B for the first round of the activity and remind them to switch roles for each round. Have them play the game according to the directions.

3 Circulate throughout the room and assist students as needed.

LAURA'S Tips

After students have played *Fishbowl Multiplication* in this setting, place the game in a math center for them to play later. Glue the directions onto the front of a manila envelope and store all game pieces inside. The envelope can serve as the "fishbowl."

FISHBOWL MULTIPLICATION
Directions
Number of Players: 2
Materials:

Fishbowl Multiplication game board
Addition and Multiplication Sentences
Bowl with 30 "fish" (paper clips or small objects)

1 Place the Addition Sentences face down in a pile. Spread out the Multiplication Sentences face up on the table.

2 Decide who is Partner A and who is Partner B. Partner A chooses an Addition Sentence and places it face up in the box at the bottom of the page. Partner A uses the numbers on the card to fill the fishbowls with equal groups of fish.

3 Partner B finds the matching Multiplication Sentence and places it face up in the other box. Partner B then counts to find the total number of fish and announces it to Partner A.

4 If Partner A agrees, remove both cards and set them aside. Put all the fish back in the bowl. If not, recount the fish.

5 Repeat steps 1 - 4, with Partner B choosing the Addition Sentence and Partner A finding the Multiplication Sentence and counting the fish.

6 Continue taking turns until all the cards have been used or the time runs out.

FISHBOWL MULTIPLICATION
Game Board

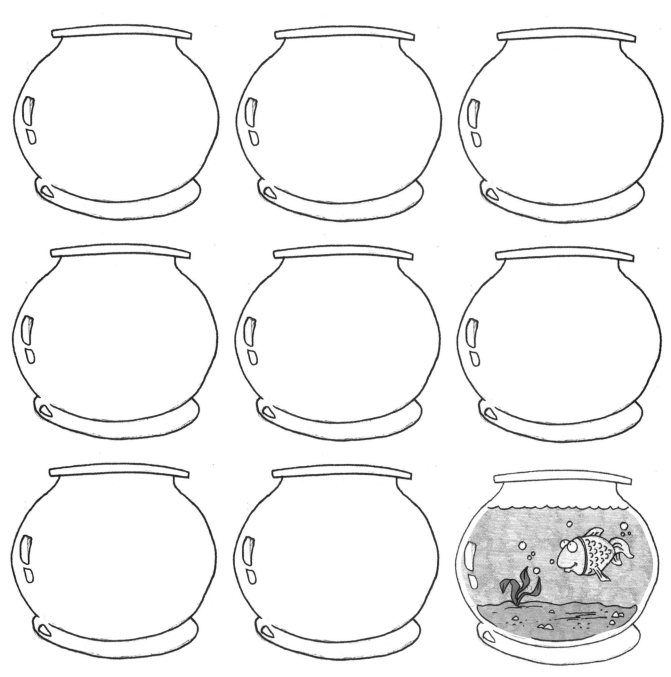

What is the total number of fish?

Addition Sentence

Multiplication Sentence

=

FISHBOWL MULTIPLICATION
Addition and Multiplication Sentences

3 + 3 + 3 + 3 + 3	5 x 3
7 + 7 + 7	3 x 7
2 + 2 + 2 + 2 + 2	5 x 2
5 + 5 + 5 + 5	4 x 5
8 + 8 + 8	3 x 8
7 + 7 + 7 + 7	4 x 7
4 + 4 + 4 + 4 + 4	5 x 4
9 + 9	2 x 9

STEP 2 — Apply Multiplication to Real-World Problems

After developing the concept of multiplication as groups of objects, it's time to show real uses of multiplication through word problems. The two activities in this step will have students act out multiplication problems, and then draw the solutions. In *Do the Math: Multiplication,* you introduce your students to multiplication word problems in a whole group setting with students acting out the solutions. The *Picture This: Multiplication* activity provides the paper-and-pencil follow-up necessary for students to transfer these real-life problem-solving strategies to more abstract applications.

Each activity is aligned with one or more Common Core Math Content and Practice Standards. The specific Standards covered are noted on the teacher information page for each activity.

- **DO THE MATH: MULTIPLICATION (PAGE 32)**
- **PICTURE THIS: MULTIPLICATION (PAGE 34)**

Do the Math: Multiplication

Students will role-play how to solve each problem using real objects and then draw the solutions on dry-erase boards.

Approximate Time:
30 minutes

Procedure

1 Display the first problem on an interactive white board or overhead projector.

> Sarah put 3 pencils on each desk in her team. If there were 4 desks, how many pencils did she use?
>
> Addition: _____ Multiplication: _____

Materials ✏️ ✂️ 📎

- 1 copy of the *Do the Math: Multiplication* problems to display (page 33)

- Items described in the four word problems: 12 pencils, 4 student desks, 15 staples, 5 papers and a bulletin board, 12 stickers and 6 papers, 20 blocks

- 1 dry-erase board and marker for each student

2 Read the problem aloud and ask a student to role-play Sarah and place 3 pencils on each of 4 different desks. Ask students how they could draw a simple sketch to show the solution. Discourage them from adding extra details that take time and distract from the solution. Have students draw their sketches of the solution on their dry-erase boards. Something similar to this drawing is all that's needed:

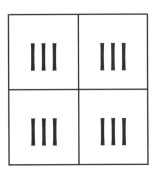

3 Next, ask students to write the addition sentence represented by this picture. They should write "3 + 3 + 3 + 3 = 12" on their dry-erase boards. Check their answers.

4 Finally, ask students to write the multiplication number sentence for this problem. The first number is the number of groups and the second number is the number of objects in each group, so this number sentence would read "4 x 3 = 12." Remind them that this picture shows 4 groups of 3 pencils.

5 Repeat these steps with the remaining *Do the Math: Multiplication* word problems.

DO THE MATH
Multiplication

Sarah put 3 pencils on each desk in her team. If there were 4 desks, how many pencils did she use?

Addition: _____ Multiplication: _____

Greg used 3 staples to put up each of the 5 papers on the bulletin board. How many staples did he use?

Addition: _____ Multiplication: _____

Mrs. Jones put 2 stickers on each paper she graded. If she graded 6 papers, how many stickers did she use?

Addition: _____ Multiplication: _____

Javier built 4 towers out of blocks. Each tower was 5 blocks tall. How many blocks in all?

Addition: _____ Multiplication: _____

Picture This: Multiplication

Common Core Standards:
3.OA.A.1, 3.OA.A.3
MP1, MP4

Picture This: Multiplication is the follow-up to *Do the Math: Multiplication*, eliminating the role-playing step. Students solve word problems by reading a problem, drawing a solution, and writing the corresponding addition and multiplication sentences.

Materials ✏️ ✂️ 📎

- 4 *Picture This: Multiplication* problems for each student

Teacher Preparation

Duplicate all the *Picture This: Multiplication* problems for each student (unless you plan to display them for the class and have the students draw their pictures on dry-erase boards or in math journals). You can download more problems at **www.lauracandler.com/mmf** and create more from the template (page 39). **ONLINE**

Approximate Time:
10 minutes per problem

LAURA'S **Tips**

Use the blank template to create additional word problems with different levels of difficulty.

Procedure

1 Seat students in pairs and give all of your students the same *Picture This: Multiplication* activity sheet as you demonstrate the procedure (or display a copy for the class).

2 Ask your students to read the problem individually and draw their solutions. Remind them to keep their drawings simple and include only the details needed to solve the problem. Check to be sure they have done this before moving on.

3 Ask partners to compare and discuss their drawings with each other. If students find errors in their work, they may correct them.

4 Ask students to write the addition and multiplication sentences to represent the problem. They should include the numeric solution that will be the same for both. When they are finished, have them compare and discuss with their partners.

5 Repeat steps 1 through 4 with the remaining problems.

PICTURE THIS: MULTIPLICATION
Problem 1

Sally put 2 flower vases on the table. She put 3 flowers in each vase. How many flowers in all?

Addition Sentence: _____

Multiplication Sentence: _____ Product: _____

PICTURE THIS: MULTIPLICATION
Problem 2

Randy had 3 baskets. Each basket had 4 apples.
How many apples in all?

Addition Sentence: _____

Multiplication Sentence: _____ Product: _____

PICTURE THIS: MULTIPLICATION
Problem 3

Jamal gave 5 pieces of candy to each of his 4 friends. How many pieces of candy did he give away in all?

Addition Sentence: _____

Multiplication Sentence: _____ Product: _____

PICTURE THIS: MULTIPLICATION
Problem 4

Tameka bought 3 boxes of crayons. Each box had 8 crayons. How many crayons did she buy?

Addition Sentence: _____

Multiplication Sentence: _____ Product: _____

PICTURE THIS: MULTIPLICATION

Addition Sentence: _____

Multiplication Sentence: _____ Product: _____

STEP 3 Represent Multiplication with Arrays

Arrays are another important multiplication tool to introduce. They are a perfect way to teach your students specific multiplication terminology. Common Core Mathematical Practice Standard MP6 emphasizes the importance of communicating precisely when discussing math concepts, and this includes development of specific mathematical vocabulary.

The first step is to teach students what is meant by the term *array*. Let them create arrays with objects and then make the transition to coloring arrays on graph paper. The last activity shows students how the multiplication table is like a huge collection of arrays organized in a meaningful way.

If materials are limited, seat students in pairs and let them work with a partner. As soon as your students master each concept, move to the next activity. Depending on your students' prior experience with multiplication concepts, ten minutes might be enough on each of the first few activities. Throughout this sequence of lessons, you'll introduce these terms: *array*, *row*, *column*, *factor*, *product*, and *commutative property*. If you skip a lesson, be sure to include any missed terms in another lesson.

Each activity is aligned with one or more Common Core Math Content and Practice Standards. The specific Standards covered are noted on the teacher information page for each activity.

- **OBJECT ARRAYS (PAGE 41)**
- **LINKING CUBE ARRAYS (PAGE 43)**
- **GRAPH PAPER ARRAYS (PAGE 45)**
- **RECTANGLE RACE (PAGE 48)**
- **MULTIPLICATION TABLE ARRAYS (PAGE 52)**

Object Arrays

Common Core Standards:
3.OA.A.1
MP4, MP6, MP7

It's easy to create an array with a collection of identical objects such as centimeter cubes, pennies, bingo chips, dried beans, plastic counters, or cereal pieces.

Approximate Time:
20 minutes

Materials ✎ ✂ 🖇

- At least 50 small identical objects for each student or pair of students
- Dry-erase boards and markers for each student

Procedure

1 Create a simple 3 x 5 array of circles on the board and explain that this is an "array," or an arrangement of objects in rows and columns.

2 Point out that the word "row" refers to each line of circles arranged horizontally across the board and the word "column" refers to each line of circles arranged vertically.

3 Ask students, "How many rows?" Then ask, "How many objects in each row?" Count to show that the total is 15 circles.

4 Explain that an array is a way of showing a multiplication sentence. Write "3 x 5 = 15" and explain that the first number describes how many rows and the second number describes how many objects in each row, or how many columns.

5 Remind students that when two numbers are added, the numbers you add are called "addends," and the answer is the "sum." Explain that with multiplication, the numbers you multiply are called "factors," and the answer is the "product."

6 Give each student a bag of identical items for creating their own arrays. If materials are limited, two students may share one bag and work together.

7 Write a new multiplication sentence on the board and have students create the array with the objects on their desk. Remind them that the first number is the number of rows and the second number is the number of objects per row. Then have them write the multiplication sentence represented by the array on their dry-erase boards so that you can verify their work.

8 Pose questions to review the terminology you have introduced (*array*, *row*, *column*, *factor*, and *product*). When you ask each question, have students respond in unison or whisper the answer to a partner.

9 Continue to write multiplication sentences on the board and allow students to create the corresponding arrays. Be sure to create some number sentences that have the numbers reversed. For example, write "4 x 6" one time and "6 x 4" the next so students can see that these two arrays have the same total number of objects.

Linking Cube Arrays

Common Core Standards:
**3.OA.A.1, 3.OA.A.4, 3.OA.B.5
MP4, MP5, MP6, MP7**

Creating arrays from linking cubes or snap cubes is the next logical step to teach arrays. If possible, use cubes that are the same size as the graph paper used in the Graph Paper Arrays lesson (¾ inch). Linking cubes are extremely effective for introducing and exploring many multiplication concepts. If your students are new to multiplication, plan to spend at least an hour on this lesson, and divide the session into two days.

Materials ✏️ ✂️ 📎

- At least 60 plastic linking cubes per student or pair (10 cubes of 6 different colors)

- Dry-erase boards and markers for each student

Approximate Time:
60 minutes (two 30-minute sessions)

Procedure

1 Give each student (or pair) at least 60 linking cubes—10 cubes of 6 different colors.

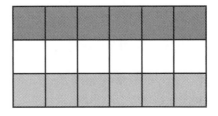

2 Write "3 x 6" on the board and ask students how they can model that math fact with their linking cubes to find the answer. They should create 3 rows of 6 cubes each as shown. Point out that even though the cubes are linked together, this is still an array because the cubes are arranged in equal rows and columns. Creating each row in a different color may help them to identify the number of rows more easily.

3 Ask students to write the multiplication sentence on their dry-erase boards, including the answer (3 x 6 = 18). Check the answers yourself or have students compare with a partner.

4 Present students with a number sentence that has an unknown quantity represented by a question mark or a blank and ask them to create an array to find the answer. Begin with an example in which you provide the factors, such as 4 x 2 = ? and ask them to use their linking cubes to create an array that models this problem. By creating an array with 4 rows of 2 blocks, they will be able to solve the problem and write "4 x 2 = 8."

5 Present additional math facts as needed to develop these concepts. As you work through each problem, review the terminology you have introduced (*array*, *row*, *column*, *factor*, and *product*). When you ask each question, have students respond aloud in unison or whisper the answer to a partner. Encourage students to use the precise vocabulary you have introduced as they respond.

6 Present a problem in which you supply one factor and the product and challenge them to find the missing factor. For example, write 5 x ___ = 20 and ask if anyone can figure out a way to find the unknown number. If they start by creating a row of five and continue to add more rows until they have 20 cubes, they will see that 4 rows of 5 were needed to find the solution. Have them rewrite the problem on their dry-erase boards with the missing factor in place (5 x 4 = 20). Present additional problems and observe their progress until everyone grasps how to find an unknown factor or an unknown product. Continue to work on using precise mathematical vocabulary during this lesson.

7 Give students pairs of math facts in which the factors are reversed. When they create these arrays, they will discover that while the total number of cubes is the same, the arrangement of rows and columns will appear different as shown.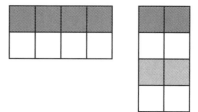

8 If you have already introduced your students to the commutative property for addition, this is a great time to introduce the concept as it relates to multiplication. Building related sets of linking cube arrays clearly shows that changing the order of the factors does not affect the product. After creating two arrays as shown in step 7, write the phrase "Commutative Property of Multiplication" on the board and write the number sentences for both arrays on the board. For the example shown in step 7, you would write "2 x 4 = 8" and "4 x 2 = 8." Challenge students to use linking cubes to create two arrays that demonstrate the commutative property of multiplication. After they have created their arrays and written their corresponding number sentences, ask them to turn to a partner and share their work.

Graph Paper Arrays

Common Core Standards:
3.OA.A.1
MP4, MP5, MP6, MP7

Another effective way to teach arrays is have students color arrays on graph paper and cut them out. This lesson works best directly following the lesson in which students build arrays with linking cubes. Use the graph paper on page 47 for this activity because the ¾-inch squares are the same size as those on the multiplication table on page 54, and they also correspond with the size of linking cubes. If your linking cubes are not the same size as the graph paper, skip the first few steps in which they trace the linking cubes and start by having them draw their arrays directly on the graph paper.

Materials ✏ ✂ 📎

- 1 sheet of graph paper (page 47) for each student
- Scissors for each student
- At least 60 linking cubes for each student that are the same size as the blocks on the graph paper
- Colored pencils, crayons, or markers

Approximate Time:
45 minutes

Procedure

1 Write "4 x 3" on the board and ask students to create an array with the linking cubes to model this math fact. Remind them to use a different color for each row of cubes and check their models when finished.

2 Ask students to place their linking cubes onto a sheet of graph paper so that the rows of cubes line up with the rows of squares. Show them how to start in the upper left corner of the paper to save room for future arrays.

3 Now ask your students to trace around their entire array to form a rectangle on the graph paper. Ask them to color the rectangle very lightly with each row being a different color to match their plastic linking cube models.

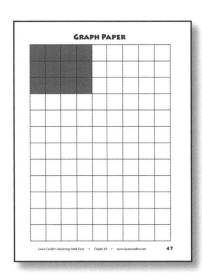

4 Discuss the terms from the previous lesson as they relate to the drawing:
- How many *rows*?
- How many *columns*?
- What are the *factors*?
- What is the *product*?

5 Have students cut around the entire array, keeping the whole rectangle intact. Ask them write the math fact and its solution in marker or with a dark pencil in the center of the rectangle. In this case, it would be 4 x 3 = 12.

6 Repeat steps 1 through 4 if your students need additional practice with concrete manipulatives.

7 Tell the students that they are going to create arrays by drawing them directly on their paper without tracing the linking cubes. Write "3 x 5" on the board and remind them that this means 3 rows of 5 squares.

8 Have them lightly shade in each of the 3 rows in different colors. Show them where to start on their paper so that they don't waste paper and will leave space available for additional problems. Have them cut out the array. Discuss the relevant terminology as it relates to their arrays.

9 Repeat step 8 at least 5 more times. Include several math facts that demonstrate the commutative property for multiplication, such as "7 x 4" and "4 x 7."

10 Have students paper clip their arrays together and put their names on the back of the stacks. Collect their arrays to use when introducing the multiplication table.

Graph Paper

Rectangle Race

Common Core Standards:
3.OA.A.1, 3.OA.C.7
MP4

Rectangle Race is a partner game in which students create arrays on a rectangular grid according to the numbers rolled on the dice and try to capture the most squares. *Rectangle Race* works best as a two-person game. This game can be played in a whole-class setting with students seated in pairs, or it can be a math center review activity. To use in a math center, glue the directions on the front of a large manila envelope and place all materials inside.

Teacher Preparation

Duplicate one *Rectangle Race* game board (page 51) for every pair of students. Make sure you have one pair of dice, two different color crayons, and a calculator or times table chart to check answers for each pair of students.

Approximate Time:
45 minutes

Materials ✏ ✂ ✁

● 1 Rectangle Race game board (page 51) for each pair of students

● 1 set of Rectangle Race game directions (page 50) to display

● 2 different color crayons for each pair of students

● 2 dice for each pair of students

● 1 calculator or times table chart for each pair of students

Introducing the game

1 Display the *Rectangle Race* directions (page 50) for the whole class and review them together.

2 Play the game against the class before allowing the students to play with a partner. Choose one color for yourself and designate one color for the class. You roll the dice first, and outline an array based on the two dice (example: rolling a 3 and a 5 gives you a 3 by 5 array of 15 squares). Lightly shade in the array, then count the squares and write that number on the array.

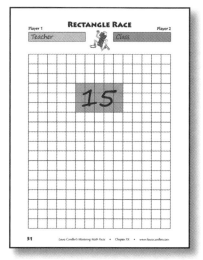

3 Tell students that the arrays can be drawn without regard to which number was rolled first. For example, if a student rolls a 3 and a 5, he

or she may draw a rectangle with 3 rows and 5 columns or one with 5 rows and 3 columns. However, arrays may not overlap.

4 Call a student to come up, roll the dice, and mark the new array on the board using the class color. Make sure the student also writes the total number of squares in the array. Then take another turn yourself. Choose a different student each time the class takes a turn.

5 Continue to play against the class until one of you rolls an array that can't be drawn on the board without overlapping. When that happens, the player loses a turn. Continue playing a few more rounds but be sure to end with both players taking an equal number of turns.

6 Count the total squares captured by each color to determine the winner.

7 After students understand how to play, allow them to play with a partner or make the game available to play in a math center.

RECTANGLE RACE

NUMBER OF PLAYERS: 2

Object of the game: To create arrays and capture the most squares on the gameboard.

MATERIALS:

- 1 Rectangle Race game board
- 2 different color crayons
- 2 dice
- Calculator or times table chart

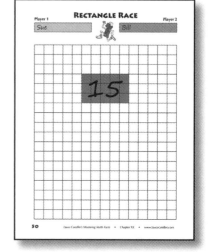

DIRECTIONS:

1 Each player chooses one color to use on the game board. Color the key at the top of the board accordingly.

2 Player 1 rolls the dice and calls out the 2 numbers. Player 1 draws and colors in a rectangular array of boxes on the game board, based on the two numbers rolled on the dice. Player 1 counts the total number of boxes in the array and writes that number in the middle. Player 2 checks the answer by entering the multiplication fact into a calculator or using a times table chart.

3 Player 2 then rolls the dice and uses a different color to draw a rectangle. He or she records the total squares in the center. Player 1 checks the multiplication fact with the calculator or chart.

4 Players continue taking turns rolling the dice and drawing rectangles. Rectangles may not overlap. If a player cannot create a rectangle in the open space, he or she loses that turn. The game ends when time runs out or the game board is filled.

5 At the end of the game, players add up their points. They score a point for each square they colored on the board. The winner is the one with the most points.

RECTANGLE RACE

Player 1

Player 2

Multiplication Table Arrays

Common Core Standards:
3.OA.A.1, 3.OA.B.5
MP4, MP5, MP6, MP7

This activity introduces the multiplication table by showing that it is based on the concept of arrays.

Teacher Preparation

Your students will need the stack of paper arrays they created in the *Graph Paper Arrays* activity. Duplicate one copy for each student of the multiplication table on page 54. When introducing this concept to your students, you'll need a way of displaying the steps to your class, either with an overhead projector and transparencies of the pages or with a document camera. If you have an interactive whiteboard, you can also demonstrate this concept using the interactive Java application on the Math Cats website called "Explore the Multiplication Table": **www.mathcats.com/explore/multiplicationtable.html**. *ONLINE*

Approximate Time:
45 minutes

Materials ✏ ✂ 🧷

- 1 completed multiplication table (page 54) for each student
- At least 6 paper arrays from the *Graph Paper Arrays* lesson for each student
- 1 dry-erase board and marker for each student

Procedure

1 Display a copy of the completed multiplication table (page 54) and explain that this table is a handy chart of all of the math facts from 1 x 1 to 9 x 9. Tell them that when we talk about all math facts together, we often refer to them as the "times tables." Explain that while the multiplication table might look confusing, it is based on patterns and is easy to read.

2 Give each student a copy of the multiplication table. Ask them to study it to see if they can discover patterns. Call on different students to share the patterns that they see. They might notice that it's like a Hundreds Board without the 10s column. They might mention that the numbers going across or down can be found by skip counting. Allow time for exploration and discussion.

3 Give students their paper arrays that they cut out previously in the *Graph Paper Arrays* lesson. Ask if they can find any patterns using those arrays and the multiplication charts. Some students may notice that if you place an array on the chart starting in the upper left corner, the number directly under the lower right corner is the total number of squares in the array.

4 If no one notices the connection between arrays and the multiplication chart, demonstrate how to place an array on the chart using the 3 x 5 paper array from the previous lesson. Ask everyone to find their own 3 x 5 rectangle and place it on their own boards as shown. Then ask them to peek under the lower right corner to see what they find—the number 15, which is the product of 3 times 5!

MULTIPLICATION TABLE

X	1	2	3	4	5	6	7	8	9
1						6	7	8	9
2						12	14	16	18
3					15	18	21	24	27
4	4	8	12	16	20	24	28	32	36
5	5	10	15	20	25	30	35	40	45
6	6	12	18	24	30	36	42	48	54
7	7	14	21	28	35	42	49	56	63
8	8	16	24	32	40	48	56	64	72
9	9	18	27	36	45	54	63	72	81

5 Have students rotate their graph paper arrays 90 degrees and again align them with the top edge and left side of their charts. When they peek under the lower right corner again, they'll notice that the answer is the same. Due to the commutative property of multiplication, 5 x 3 = 3 x 5, so it doesn't matter which way they turn their paper arrays as long as they align them properly with the left and top edges.

6 Allow time for students to continue exploring with their paper arrays until they realize that this relationship holds true for all of their arrays. Walk around and check to be sure all students are placing their arrays on the grid properly; remind them that the X in the upper corner should be visible. Make sure they don't cover the factors going across the top and down the left side.

7 Demonstrate how to find the product of two factors using the multiplication table alone, without the paper arrays. Point out that the bold numbers across the top and down the left side are the factors in a multiplication problem. The X in the upper left corner reminds them to multiply, and the products are found in the middle of the chart. To find the answer to any multiplication fact, they locate the factors on top and side and where that row and column intersect. Use an example like 2 x 4 and show how to find the product.

8 Present multiplication facts to your class, one at a time, and ask them to find the product of each math problem using their charts. Ask them to write the multiplication fact and the product on their dry-erase boards as a number sentence (example: 2 x 4 = 8). Continue to present problems to students until they are comfortable using the multiplication table.

MULTIPLICATION TABLE

X	1	2	3	4	5	6	7	8	9
1	1	2	3	4	5	6	7	8	9
2	2	4	6	8	10	12	14	16	18
3	3	6	9	12	15	18	21	24	27
4	4	8	12	16	20	24	28	32	36
5	5	10	15	20	25	30	35	40	45
6	6	12	18	24	30	36	42	48	54
7	7	14	21	28	35	42	49	56	63
8	8	16	24	32	40	48	56	64	72
9	9	18	27	36	45	54	63	72	81

STEP 4 Teach All Multiplication Facts

After students have explored multiplication through hands-on activities and arrays, it's time to teach them all of the multiplication facts in a sequential manner. This process is likely to take several weeks; students will learn the math facts, then need time to practice them to build fluency. Explain that in order to be successful in math, they must memorize the times tables and know them quickly and accurately. Skip-counting on fingers is fine at the beginning, but in the end, they must know them by heart.

Introduce Each Math Fact Individually

When students first learn the times tables, they need to be introduced to each set of multiplication facts individually. As you introduce each fact, students need hands-on experiences, but they also need to record the products and begin to memorize them. As the students work through the times tables using manipulatives, it's helpful to have them record their findings in three places:

- **MULTIPLICATION FACTS TO GO BOOKLET (PAGE 60)**
- **BLANK MULTIPLICATION TABLE (PAGE 59)**
- **INDIVIDUAL SET OF FLASH CARDS** ONLINE

You may want to begin by introducing 0 and 1 on the same day because these two are easy to grasp. Then spend at least one day introducing each of the remaining facts. Use your math curriculum to decide whether to introduce all facts up to 12 or to stop with 9 or 10. The *Introducing Math Facts* activity gives a sample lesson for introducing each new multiplication fact.

This activity is aligned with several Common Core Math Content and Practice Standards. The specific Standards covered are noted on the teacher information page for the activity.

LAURA'S Tips

You may want to introduce the multiplication facts in order from easiest to learn to most difficult. Many teachers use this sequence: 0, 1, 2, 5, 10, 3, 4, 6, 7, 8, 9, 11, 12

Introducing Math Facts

Common Core Standards:
3.0A.C.7
MP4, MP5, MP7

Teacher Preparation

Create a *Multiplication Facts to Go* booklet for each student. Duplicate the pattern (page 60) and cut the sections apart. Stack the pages in order and staple them along the left edge to make a small booklet.

Print one set of flash cards for each student, using the patterns that can be downloaded at **www.lauracandler.com/mmf**. It's helpful to use different colors of paper or card stock for each number. For example, print the 2s on pink paper, the 3s on yellow paper, and so on. Students can store all their flash cards together but can easily separate them when they want to study a particular set of math facts.

Approximate Time:
1 to 2 days per math fact

Materials ✏ ✂ 📎

- *Multiplication Facts to Go* booklet (page 60)
- 100 small identical objects for creating arrays (dried beans, counters, etc.)
- Blank multiplication table (page 59) for each student
- Individual sets of flash cards printed on construction paper or card stock 🖱**ONLINE**
- Plastic zip bags for flash card storage
- Calculator
- Mini Math Check Sheets (pages 154 – 156)

Math Facts 0 and 1 Procedure

1 Give each student a handful of manipulatives for creating arrays. Write 0 x 4 on the board and ask them to create an array to show that math fact. They will probably seem confused. Explain that you can't show that math fact with manipulatives because it means zero rows of 4 objects which is still zero. Demonstrate with a few more examples.

2 Give each student his or her own set of flash cards for the 0 facts. Explain that just as with addition number sentences, we often write multiplication sentences vertically. So 4 x 0 can be written with the 4 above the 0 as they see it on the flash cards. Have them flip the cards face down and write their names at the top of each card and the answer near the bottom. Remind them to write very lightly so the answer is not visible from the other side.

3 To begin introducing the 1s, write a problem such as "1 x 4" on the board and ask students to create an array to model what this means. This would be 1 row of 4 objects so the product is just 4. Repeat with several examples until they see that the product of any number and 1 is that number.

4 Give each student a blank *Multiplication to Go* booklet, and tell them that this is where they will keep a record of all the times tables. Have them write their names on the covers. Then ask them to fill in the products on the 1s page, which, they will discover, is a very easy task!

5 Give each student a blank copy of the multiplication table to begin filling out. Show them how to fill in all of the products in the top row across from the number 1, and in the column down the left side under the number 1.

6 Finally, give out the flash cards for the 1s. As with the 0 flash cards, have students write their names and the product on the back of each card. All flashcards should be stored together in a plastic zip bag labeled with their name. You may want to collect their bags, booklets, and multiplication tables to keep at school.

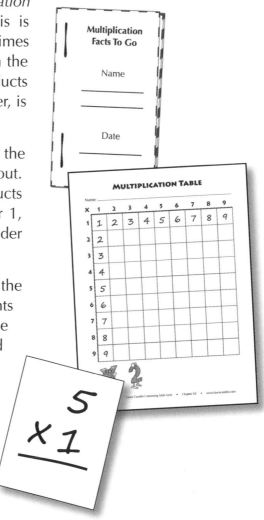

Math Facts 2 through 12 Procedure

LAURA'S Tips

If this is the first time your students have been introduced to the math facts, plan to spend at least one day introducing each math fact from 2 through 10 or 12.

1 For each set, present the math facts in order starting with 2 times the number. Have students use manipulatives, as in the lessons for teaching the 0s and 1s, to build increasingly larger arrays and record the products on their multiplication charts. For example, remind them that they already know what 1 x 2 is, but ask if they can show you 2 x 2. When they create the array and find the product to be 4, have them record this on the row for the 2s on their multiplication table. Don't use the booklet or the flash cards at this point. Ask them to find and record the products for 2 x 3, 2 x 4, and so on. They will soon notice that the products follow a predictable pattern, which is just like skip-counting in addition.

2 After they complete the row for the day's math fact, have them complete the corresponding page in their *Multiplication Facts to Go* booklet. Check their written responses after they are recorded to ensure that their answers are correct.

3 Give them the set of flash cards for that day's math fact. Have them write their names and the answers on the back of each flash card. Because the problems will get increasingly difficult, encourage students to check their answers with a calculator to be sure they are correct.

4 Provide at least 10 minutes for your students to study the day's math facts with their flash cards, preferably with a partner, because partner practice is more effective than studying alone. After 5 minutes, remind them to switch roles, whether or not they have completed the set of flash cards.

5 Wrap up the lesson by giving students an individual written quiz using one of the *Mini Math Check* sheets (pages 154 - 156). Tell students they must answer the problems in the order they are presented on the page, from top to bottom. Give an appropriate amount of time for them to complete this task; 1 or 2 minutes should be sufficient.

LAURA'S Tips

Show your students how to study the math facts with flash cards. Demonstrate methods for individual study and for working with a partner, using the directions on pages 175 - 176 in Chapter 4. These directions are appropriate for a mixed deck of flash cards or a single set of flash cards for one number.

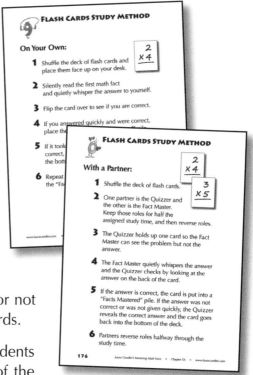

MULTIPLICATION TABLE

Name: _____

X	1	2	3	4	5	6	7	8	9
1									
2									
3									
4									
5									
6									
7									
8									
9									

MULTIPLICATION FACTS TO GO BOOKLET

Multiplication Facts To Go

Name _____

Date _____

1s

1 x 1 = _____
1 x 2 = _____
1 x 3 = _____
1 x 4 = _____
1 x 5 = _____
1 x 6 = _____
1 x 7 = _____
1 x 8 = _____
1 x 9 = _____

2s

2 x 1 = _____
2 x 2 = _____
2 x 3 = _____
2 x 4 = _____
2 x 5 = _____
2 x 6 = _____
2 x 7 = _____
2 x 8 = _____
2 x 9 = _____

3s

3 x 1 = _____
3 x 2 = _____
3 x 3 = _____
3 x 4 = _____
3 x 5 = _____
3 x 6 = _____
3 x 7 = _____
3 x 8 = _____
3 x 9 = _____

4s

4 x 1 = _____
4 x 2 = _____
4 x 3 = _____
4 x 4 = _____
4 x 5 = _____
4 x 6 = _____
4 x 7 = _____
4 x 8 = _____
4 x 9 = _____

5s

5 x 1 = _____
5 x 2 = _____
5 x 3 = _____
5 x 4 = _____
5 x 5 = _____
5 x 6 = _____
5 x 7 = _____
5 x 8 = _____
5 x 9 = _____

6s

6 x 1 = _____
6 x 2 = _____
6 x 3 = _____
6 x 4 = _____
6 x 5 = _____
6 x 6 = _____
6 x 7 = _____
6 x 8 = _____
6 x 9 = _____

7s

7 x 1 = _____
7 x 2 = _____
7 x 3 = _____
7 x 4 = _____
7 x 5 = _____
7 x 6 = _____
7 x 7 = _____
7 x 8 = _____
7 x 9 = _____

8s

8 x 1 = _____
8 x 2 = _____
8 x 3 = _____
8 x 4 = _____
8 x 5 = _____
8 x 6 = _____
8 x 7 = _____
8 x 8 = _____
8 x 9 = _____

9s

9 x 1 = _____
9 x 2 = _____
9 x 3 = _____
9 x 4 = _____
9 x 5 = _____
9 x 6 = _____
9 x 7 = _____
9 x 8 = _____
9 x 9 = _____

STEP 5 Connect Multiplication and Division

Division can be introduced as soon as students fully grasp multiplication concepts. There is no need to wait until your students have memorized all of the multiplication tables; speed and fluency can be developed later. These activities are listed in the order in which you might want to teach them, but you can adapt them or change the sequence as needed for your class.

Each activity is aligned with one or more Common Core Math Content and Practice Standards. The specific Standards covered are noted on the teacher information page for each activity.

- **Fair Shares Exploration (page 62)**
- **Fishbowl Division - No Remainders (page 64)**
- **Fishbowl Division - With Remainders (page 67)**
- **Division Mix-Up (page 74)**
- **Math Fact Families (page 77)**

Fair Shares Exploration

Common Core Standards:
3.OA.A.2
MP4

This activity is a fun introduction to division concepts. Because it's more of an exploratory activity than an instructional one, there's no need to introduce division terminology at this point unless this is the only hands-on division lesson you plan to teach

Materials ✏ ✂ ✂

- One plastic zip bag of about 20 to 30 small items for each team

Teacher Preparation

Prepare one bag of 20 to 30 small items for each team of 3 to 5 students. Items that work well for this activity include pennies, paper clips, dried beans, bingo chips, etc.

Approximate Time:
10 minutes

Procedure

1 Create teams of 3 to 5 students. Give each team one bag of small items.

2 Before students are allowed to open the bag, ask them to discuss how they plan to share the items in a fair way. Discuss the term "fair" as each student having an equal number of pieces. Don't worry about introducing the term "division" at this point although some students may use it. Ask students to raise their hands when they have decided on a strategy. For example, they might say that one student will dump out all the pieces and start giving one to each student in turn until they run out.

3 As the hands go up in each team, walk over and ask them for a quick description of what they plan to do. Even if you don't think their idea will work, let them try it out and check back later. Ask them to raise their hands again when they are finished.

4 Ask team members to record the results in any fashion they would like. They should include the total number of items, the number of groups, the number in each group, and any items left over.

5 If time permits, ask each team to think of a different strategy for sharing the items. They might say that they will dump out the items and count all the pieces first, then place the items in groups according to the number of team members. Have them compare their results to the results they obtained the first time.

6 When the groups are finished with the activity, explain that division is the process of placing items in equal groups or cutting something so that all parts are equal in size. In some cases, the groups will end up the same size with no extra items. In other cases, the groups will be the same size but there will be a "remainder" or left over amount.

7 If you are planning to use the Fishbowl Division activity (page 64), there's no need to introduce specific division terminology during the Fair Shares exploration. However, if you are not planning to engage your students in additional hands-on division activities, introduce the terms *quotient*, *dividend*, *divisor*, and *remainder*.

Fishbowl Division

Common Core Standards:
**3.OA.A.2, 3.OA.A.4, 3.OA.B.6
MP4, MP6**

Fishbowl Division is a teacher-guided activity to help students develop the concept of division. If you used the *Fishbowl Multiplication* activity, you could start this lesson by having students play a few rounds of that game. *Fishbowl Division* is similar, but there are no student directions. This activity must be led by the teacher in order to make sure students grasp each concept as it is introduced. You can use fish-shaped crackers or any small manipulative such as plastic chips or unit cubes to represent the fish.

There are two variations of *Fishbowl Division*. In *Fishbowl Division 1*, the problems do not result in remainders. In *Fishbowl Division 2*, students will have some fish left over, which they will record as a remainder. It's best if you can display the activity using a document camera or on an interactive whiteboard so students can follow along as you demonstrate.

Materials ✏️ ✂️ 🧵

- *Fishbowl Division 1 and 2* "placemats" (pages 68 and 71), 1 for each student or pair of students

- Bowl with about 100 small "fish" (goldfish crackers, unit cubes, bingo chips, paper clips, etc.)

- Dry-erase board and marker

Teacher Preparation

Decide whether you want students to do this activity alone or in pairs. On sturdy card stock or construction paper, duplicate and laminate both versions of the Fishbowl Division "placemat" for each person or pair of students.

LAURA'S Tips

You can also print the two placemat versions back-to-back on one sheet of card stock and laminate that. If you can't laminate it, slide them back-to-back into a clear page protector so students can write on them with dry-erase markers.

Fishbowl Division 1 (No Remainders) Procedure

The directions here are for individuals; if students work with a partner, have them take turns on each step.

1 • REVIEW MULTIPLICATION

Give each student a copy of the Fishbowl Division 1 (no remainders) placemat (page 68). Begin by asking students to place 5 fish into each of 3 circles on their boards. Ask for a student to express this with a multiplication sentence. Write "3 x 5 = 15" on the board and remind them that this number sentence stands for "3 groups of 5 equals 15."

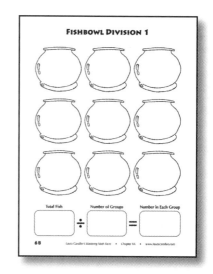

2 • INTRODUCE DIVISION AS THE INVERSE OF MULTIPLICATION

Tell students that in the same way that addition has an opposite operation (subtraction), multiplication has an opposite operation, which is division. Division reverses or "undoes" multiplication and is therefore known as the "inverse" of multiplication. With multiplication, you know the number of groups and the number in each group and have to find the total. With division, you are given the total to start with and one of the factors (such as the number of groups) and you have to find out the other factor.

3 • DEMONSTRATE DIVISION

Gather all 15 fish from step 1 into a pile and explain that with a division problem you would be given the total number (15) and one other number such as the number of groups (3). Then you would divide the total number equally into those groups and record the number in each group as the solution to the division problem. Show them how to do this by dividing the 15 fish into 3 groups by placing 5 fish into each of 3 circles. On the bottom of the *Fishbowl Division* sheet, record 15 ÷ 3 = 5.

4 • GUIDED DIVISION PRACTICE

Ask students to count out 8 fish. Have them write "8" in the "Total Number" box. Then ask them to divide the 8 fish into 4 equal groups by placing 2 fish into each of 4 circles. Show them how to record 8 ÷ 4 = 2 in the workspace at the bottom of the sheet. Repeat with additional division problems as needed, always stating the total number of fish first and then the number of fish in each group. For each problem, have students record the division number sentence in the boxes at the bottom of the sheet. Walk around and check to be sure they are following directions and recording the number sentences properly.

5 • INTRODUCE DIVISION TERMINOLOGY

A Display *Division Number Sentences* (page 69) to introduce division terminology.

B Explain that there are two ways people commonly write division problems. The first way is the horizontal method they've been using on the bottom of the Fishbowl Division placemat. Cover the bottom of the sheet so they can focus on the terminology at the top. Introduce the terms *dividend, divisor,* and *quotient.*

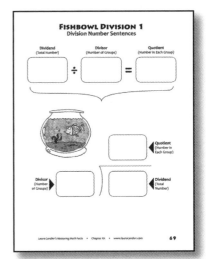

C Switch to the unlabeled template (page 70) and write in a new number sentence. Ask them to identify the dividend, divisor, and quotient.

D When they are comfortable with those terms, switch back to the labeled sheet and uncover the bottom of the page to show the other method, which involves writing the total number inside a little box, or "house." Review the proper locations of the dividend, divisor, and quotient.

E Write one division problem at a time on the board in one of the two formats, and have students write the other format on their dry-erase boards for you to check.

6 • UNDERSTAND DIVISION AS AN UNKNOWN FACTOR PROBLEM

Explain that because division is the inverse or opposite of multiplication, you can find the quotient by thinking about the related multiplication problem. Use the example in step 4 (8 ÷ 4), and explain to students that they can find the quotient by thinking, "4 x ? = 8," or "4 times what equals 8?" Students who know the times tables will know that the unknown factor is 2, which can also be observed in the arrangement of 8 fish divided into 4 groups with 2 in each group. Pose additional problems and have students first write the problem as an unknown factor multiplication problem and then solve it without using the manipulatives. Finally, have them check their answers by working out each problem on the Fishbowl Division mat.

Fishbowl Division 2 (Understanding Remainders) Procedure

1 • INTRODUCE DIVISION WITH REMAINDERS

When your students are comfortable dividing without remainders, introduce division with remainders. Use the *Fishbowl Division 2* placemat, *Understanding Remainders*. Ask students to count out 14 fish and divide them into 4 groups. Wait to see what happens. The students will notice that there are 2 extra fish. Some students may try to place those last two fish into other groups, but remind them that division means *equal* groups. Instead, have them place the extras to the side and tell them that the number of leftover fish is called the "remainder." Show them how to write the number sentence for this problem: 14 ÷ 4 = 3 r 2

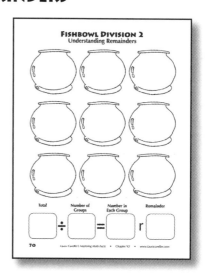

2 • GUIDED DIVISION PRACTICE WITH REMAINDERS

Continue giving your students division problems by stating the total number of fish and the number of groups. Allow time for students to work out their answers with the fish. Sometimes give them a problem that has no remainder and explain that they can leave the remainder box empty or record a zero there.

3 • MODEL DIVISION SENTENCES WITH REMAINDERS

Display the *Fishbowl Division 2 Number Sentences* pages (pages 72 and 73) and model both ways to review the previously-taught terminology. Be sure to show how to represent the remainder in number sentences.

FISHBOWL DIVISION 1

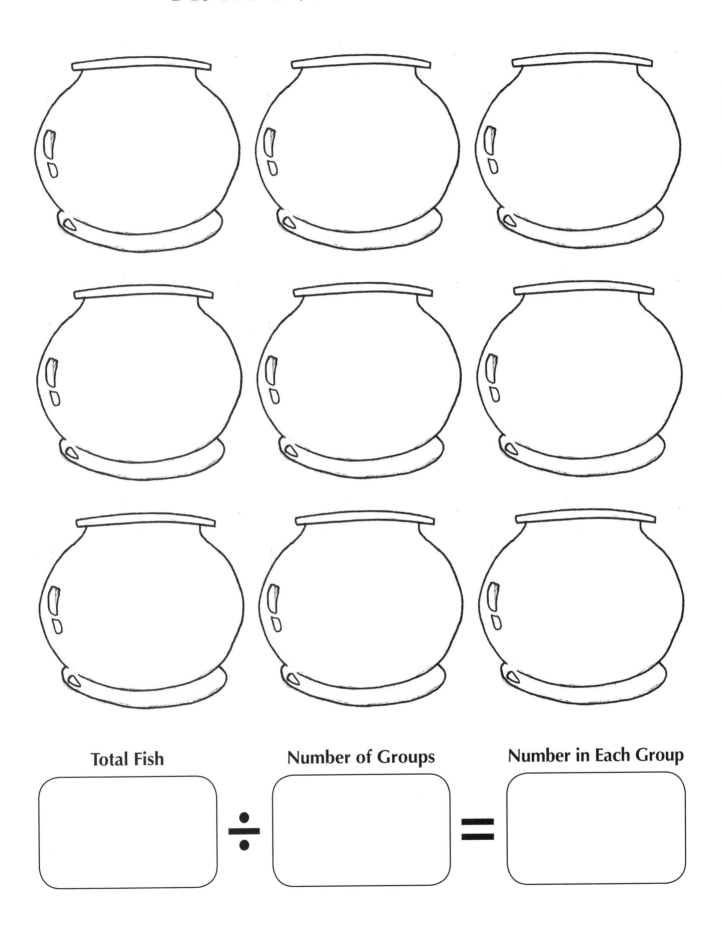

Total Fish

÷

Number of Groups

=

Number in Each Group

FISHBOWL DIVISION 1
Division Number Sentences

Dividend
(Total Number)

Divisor
(Number of Groups)

Quotient
(Number in Each Group)

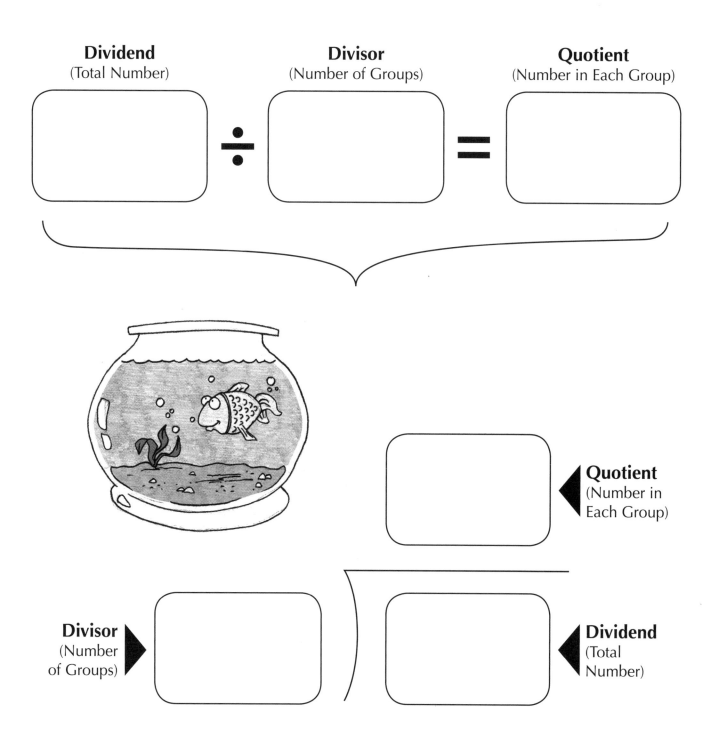

Quotient
(Number in
Each Group)

Divisor
(Number
of Groups)

Dividend
(Total
Number)

FISHBOWL DIVISION 1
Division Number Sentences

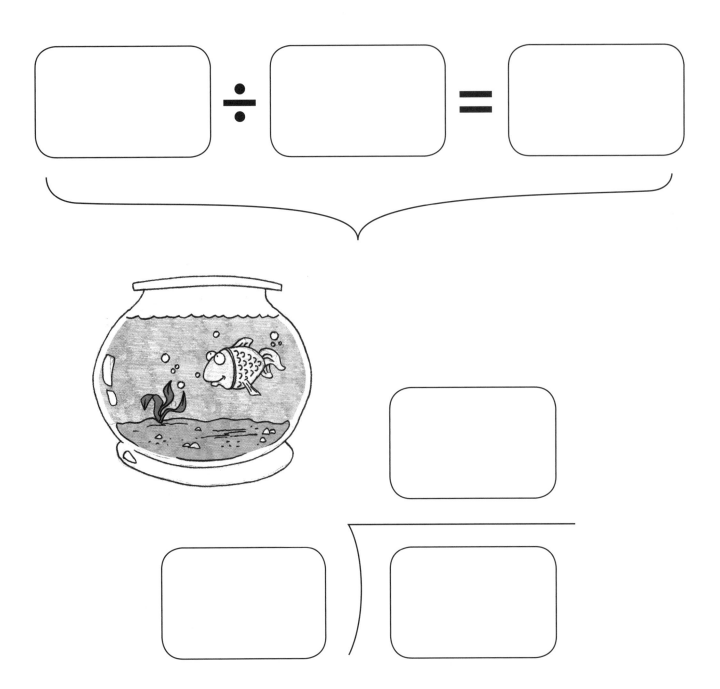

FISHBOWL DIVISION 2
Understanding Remainders

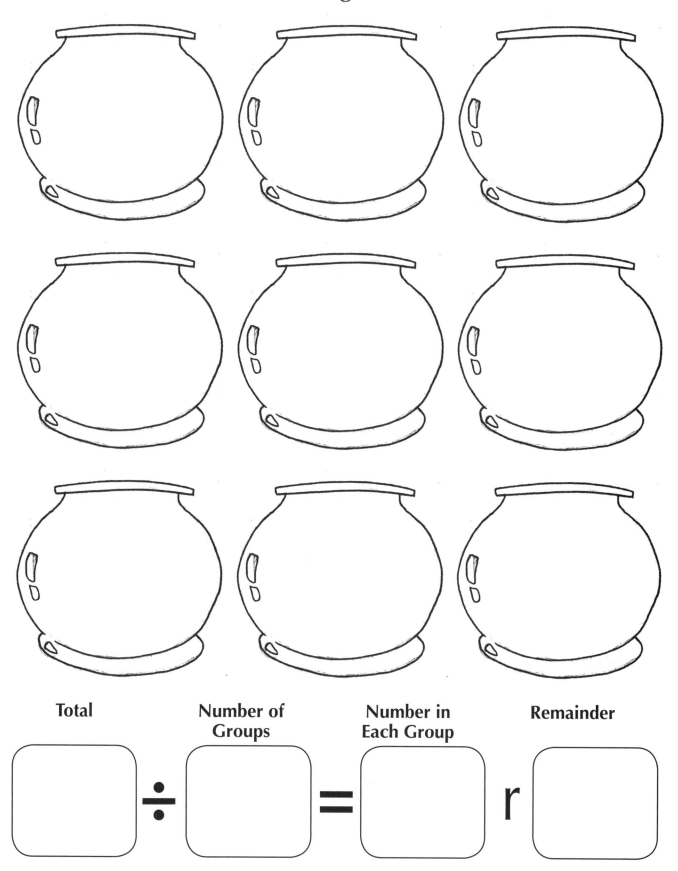

Total		Number of Groups		Number in Each Group		Remainder
☐	÷	☐	=	☐	r	☐

FISHBOWL DIVISION 2
Understanding Remainders
Division Number Sentences

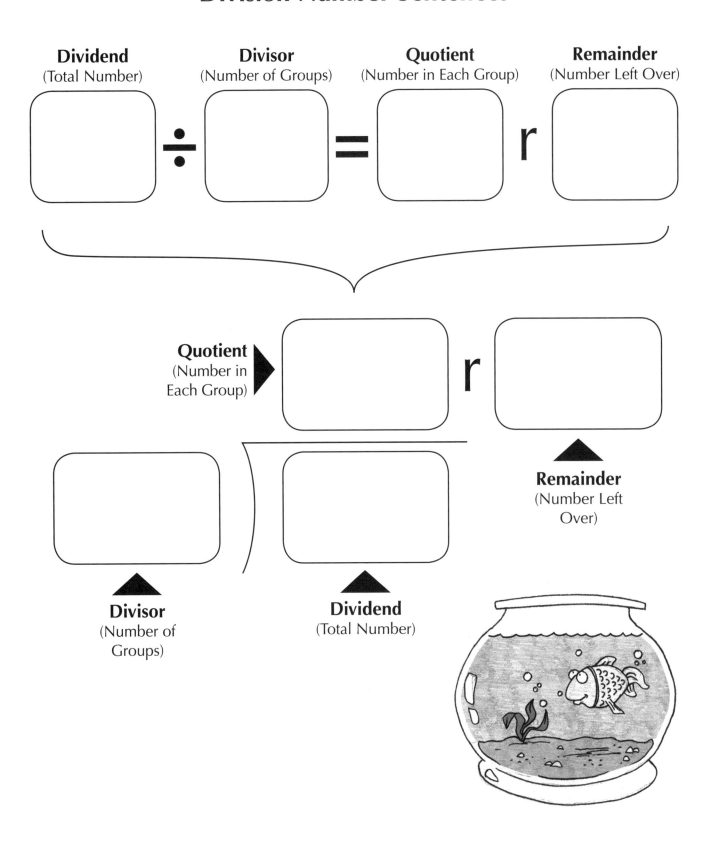

Dividend
(Total Number)

Divisor
(Number of Groups)

Quotient
(Number in Each Group)

Remainder
(Number Left Over)

÷ = r

Quotient
(Number in Each Group)

r

Remainder
(Number Left Over)

Divisor
(Number of Groups)

Dividend
(Total Number)

FISHBOWL DIVISION 2
Understanding Remainders
Division Number Sentences

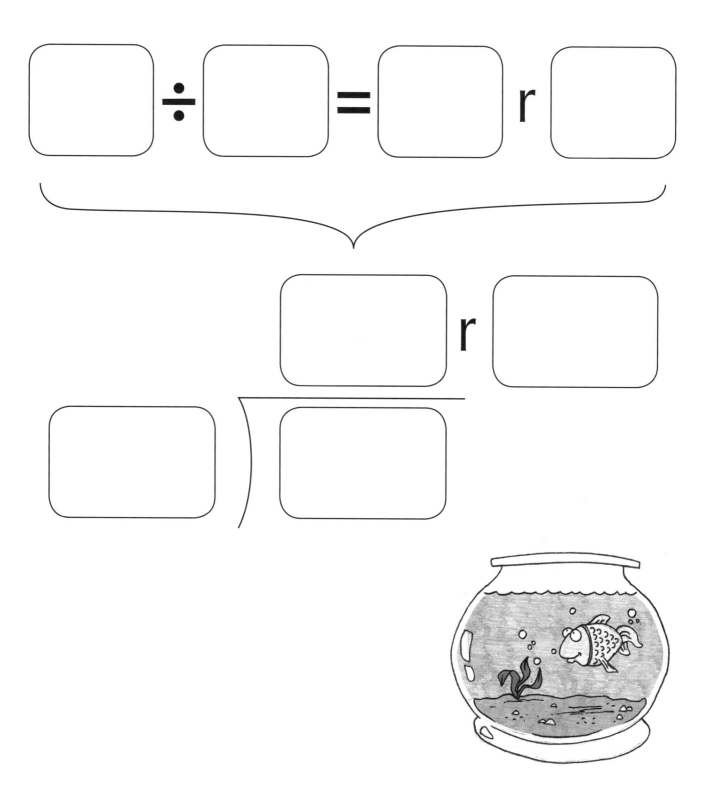

Division Mix-Up

Common Core Standards:
3.OA.A.2
MP4, MP6

Division Mix-Up is an excellent follow-up to the *Fishbowl Division* activity, and it gets students up and moving around the room. It's also a great review for older students who need a short refresher on division terminology but not the hands-on experience provided by the *Fishbowl Division* game. In this activity, students mix around the room and form groups as you write the corresponding division problem on the Recording Board.

Procedure

1 Display the *Division Mix-up Recording Board* (page 76). Write the total number of students as the dividend on the board. For this example, assume 25 for the class size.

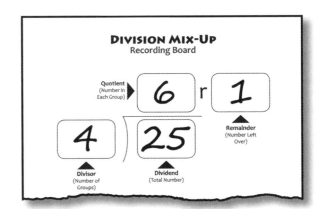

2 Call out "Mix!" and students move quietly around the room.

3 Say "Freeze!" and students stop in place.

4 Announce the group size and have students gather in groups of that size. For this example, 4 is the group size (the divisor).

5 Create a "left over" area at the front of the room for students who don't fit into any of the equal groups, and ask any extra students to join you in this area.

6 Say "We have divided into groups of 4 students. How many whole groups do we have?" Have each group count off as you point to them. In the example with 25 students in groups of 4, there will be 6 whole groups. Write the number of groups as the quotient.

7 Check to see how many students are left over. Ask, "Where shall we record these students?" Hopefully, they will suggest recording them as the remainder.

8 Tell the class that each student can only part of a remainder once in the activity. This will encourage students to include all class members and not leave anyone out.

9 Repeat the activity using the same dividend (class size) and different divisors (group sizes). To make the activity more challenging, remove the terms *divisor*, *dividend*, *quotient*, and *remainder*. After you record the complete problem on the board or recording sheet, ask students to identify the parts of the problem correctly.

LAURA'S Tips

To encourage even more active engagement, have students bring individual dry-erase boards with them and solve the division problem for each round before you record it on the board.

DIVISION MIX-UP
Recording Board

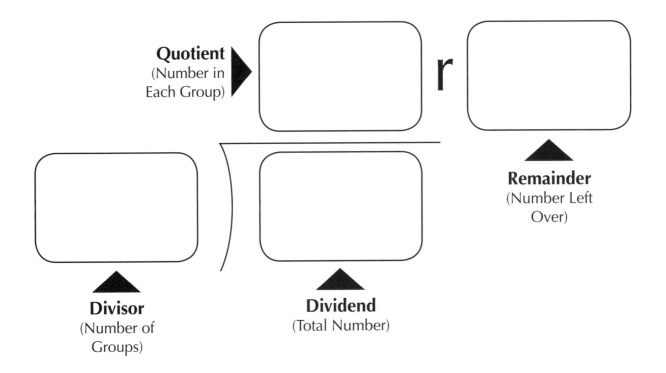

Quotient
(Number in
Each Group)

r

Remainder
(Number Left
Over)

Divisor
(Number of
Groups)

Dividend
(Total Number)

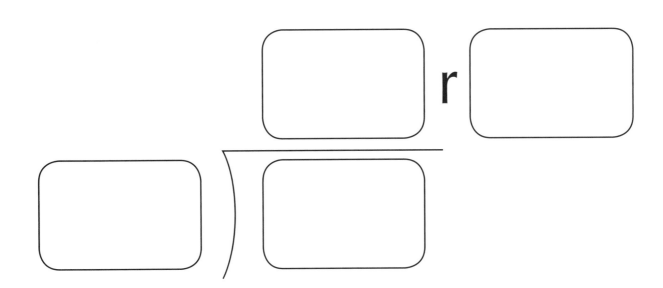

r

Math Fact Families

Common Core Standards:
3.OA.B.5, 3.OA.C.7
MP4, MP6, MP7

To be sure students understand the relationship between multiplication and division, introduce the concept of Math Fact Families. Math Fact Families are sets of multiplication and division facts that are related to each other inversely. For example, the following four math facts would be considered a math fact family:

- $3 \times 8 = 24$
- $8 \times 3 = 24$
- $24 \div 8 = 3$
- $24 \div 3 = 8$

Materials ✎ ✂ ✃

- At least 30 small identical objects per person or pair of students
- Dry-erase boards and markers for each student
- Both *Math Fact Families* printables for each student (pages 79 - 80)

Whole-Class Lesson Procedure

1 Begin by having students explore this concept with small objects like plastic chips, paper clips, cereal pieces, or math manipulatives. Give each student at least 30 objects to use for this part of the lesson.

2 Model the lesson by displaying 10 small objects in a 5 x 2 array and asking them to form this array on their own desks. Ask students to write one multiplication or division math fact on their individual dry-erase boards to represent the array, and place their boards face down when ready.

3 Now ask students to show you their boards, and record their responses until you have all four variations on the board (5 x 2 = 10, 2 x 5 = 10, 10 ÷ 5 = 2, 10 ÷ 2 = 5). Explain that these four math facts form a "fact family" because they are all ways to represent that particular array. Review the terms *factor, multiple, divisor, dividend,* and *quotient* as they relate to the math facts. This would also be a good time to review the commutative property of multiplication. Point out that the two multiplication facts in the family are easy to find because the commutative property says that reversing the order of the factors will not affect the product. Then ask them if they think there's a commutative property for division. Hopefully, they will realize that this property does not hold true for division—the larger number must always be written first.

4 Ask students to count out 24 objects and place them in a 4 x 6 array. Ask them to write all four math facts that relate to this array. Remind them that even though 8 x 3 = 24 is a number sentence that includes the correct product, this math fact does not describe the given array. The only acceptable variations are: 6 x 4 = 24, 4 x 6 = 24, 24 ÷ 4 = 6, and 24 ÷ 6 = 4.

5 Continue practicing with additional arrays of objects until everyone is able to find the four equations that make up a fact family for any given array.

6 Use the *Math Fact Families* printables (pages 79 and 80) for further practice and to assess understanding.

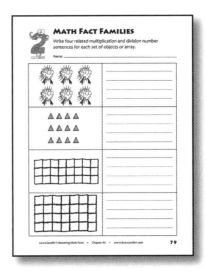

MATH FACT FAMILIES

Write four related multiplication and division number sentences for each set of objects or array.

Name: _____

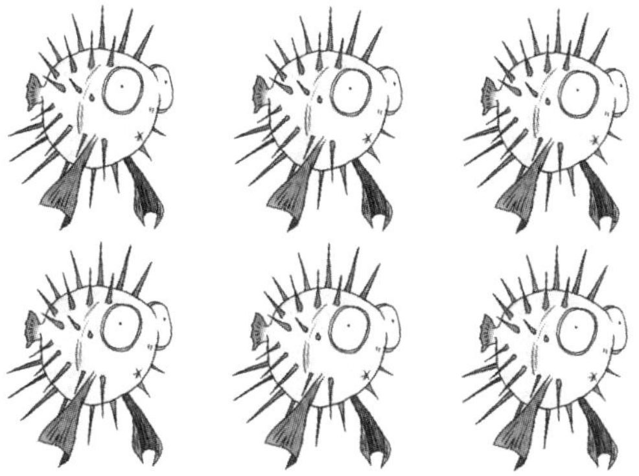

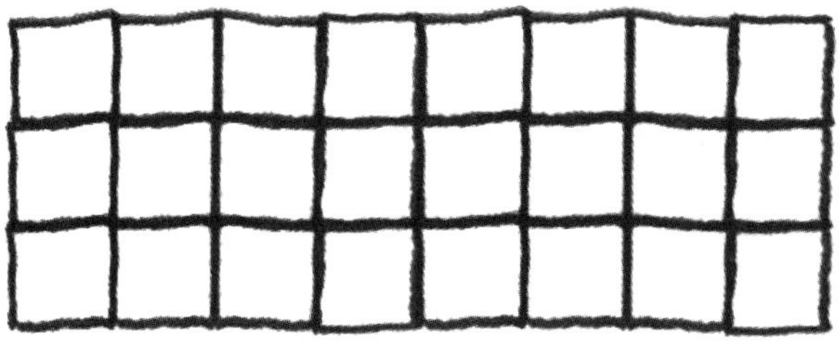

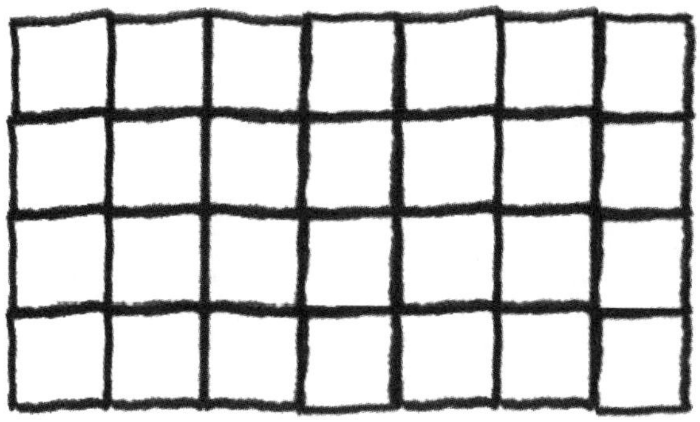

MATH FACT FAMILIES

Write four related multiplication and division number sentences for each set of objects or array.

Name: _____

<table>
<tr><td>

★ ★ ★ ★
★ ★ ★ ★

</td><td>

_____
_____
_____
_____
_____

</td></tr>
<tr><td>

☺ ☺ ☺ ☺ ☺ ☺
☺ ☺ ☺ ☺ ☺ ☺
☺ ☺ ☺ ☺ ☺ ☺
☺ ☺ ☺ ☺ ☺ ☺

</td><td>

_____
_____
_____
_____
_____

</td></tr>
<tr><td>

(grid array)

</td><td>

_____
_____
_____
_____
_____

</td></tr>
<tr><td>

(grid array)

</td><td>

_____
_____
_____
_____
_____

</td></tr>
</table>

80   *Laura Candler's Mastering Math Facts*   •   Chapter 1   •   www.lauracandler.com

STEP 6 Apply Division to Real-World Problems

After developing the concept of division as creating equal groups, it's time to share real uses of division through word problems. If you haven't done the *Fair Shares Exploration* (page 62), you may want to do that activity now. Then use *Do the Math: Division*, which allows you to introduce your students to division word problems in a whole-group setting with students acting out the solutions. Students often need guidance with how to represent a math word problem with pictures to help them solve it, so if your students are struggling with these concepts, be sure to spend adequate time on this activity. *Picture This: Division* and *Picture This: Division Challenge* provide the paper-and-pencil follow up necessary for students to transfer these real-life problem-solving strategies to more abstract applications.

Each activity is aligned with one or more Common Core Math Content and Practice Standards. The specific Standards covered are noted on the teacher information page for each activity.

- **DO THE MATH: DIVISION 1 AND 2 (PAGE 82)**
- **PICTURE THIS: DIVISION (PAGE 85)**

Do the Math: Division 1 and 2

Students will role-play how to solve each problem using real objects and draw the solutions on dry-erase boards. *Do the Math: Division 1* problems do not have remainders. *Do the Math: Division 2* problems have solutions with remainders.

Approximate Time:
30 minutes

Procedure

Decide whether you want to use the problems without remainders—*Do the Math: Division 1*—or the ones with remainders—*Do the Math: Division 2*. Follow the same directions for each set of problems.

These directions are for *Do the Math: Division 2.*

Materials ✏ ✂ 📎

- *Do the Math: Division 1* and *Do the Math Division 2* problems (pages 83 and 84)

- Items described in the word problems: 12 pencils, 25 paper clips, 20 cookies (real or paper from the patterns on page 21), box of 24 crayons

- Dry-erase boards and markers for each student

1 Display the first problem in *Do the Math: Division 2* (page 84) on an interactive whiteboard or overhead projector.

2 Read the problem aloud. Have a student role-play Julio and give him/her 10 pencils. Ask for three more volunteers to role-play Julio's friends. Ask

> Julio bought a pack of 10 pencils and gave each of his 3 friends the same number. He did not keep any for himself. How many pencils did each friend get?
>
> Division Number Sentence _____

Julio to share his pencils with the other students. Then have the friends show how many pencils they have and how many are left over.

3 Ask the students to draw something on their dry-erase boards that represents the problem. Discourage them from adding extra details that take time and distract from the solution.

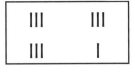

4 Have students write the division number sentence represented by this picture. Demonstrate how to write "10 ÷ 3 = 3 r 1" and have them copy it on their dry-erase boards. Review the division terms you introduced earlier.

5 Repeat these steps with the remaining word problems.

DO THE MATH
Division 1

Julio bought a pack of 10 pencils and gave each of his 5 friends the same number. He did not keep any for himself. How many pencils did each friend get?

Division Number Sentence _____

Mrs. Tevo passed 20 paper clips out to the members of a team. Each student received 4 paper clips. How many members were on the team?

Division Number Sentence _____

Sally brought in a dozen cookies and shared them equally between herself and two other students. How many cookies did each of them get?

Division Number Sentence _____

Barry gave every team member the same number of crayons. If there were 24 crayons in the box and 4 students, how many crayons did each person get?

Division Number Sentence _____

DO THE MATH
Division 2

Julio bought a pack of 10 pencils and gave each of his 3 friends the same number. He did not keep any for himself. How many pencils did each friend get?

Division Number Sentence _____

Mrs. Tevo passed 25 paper clips out to the members of a team. Each student received 4 paper clips. How many members were on the team?

Division Number Sentence _____

Sally brought in 20 cookies and shared them equally between herself and eight other students. How many cookies did each of them get?

Division Number Sentence _____

Barry gave every team member the same number of crayons. If there were 24 crayons in the box and 5 students, how many crayons did each person get?

Division Number Sentence _____

Picture This: Division

Common Core Standards:
3.OA.A.2, 3.OA.A.3
MP1, MP4

Picture This: Division eliminates the role-playing step of *Do the Math: Division*. This activity is the division version of *Picture This: Multiplication*. *Picture This: Division* problems (pages 87 - 90) do not have remainders; *Picture This: Division Challenge* problems (pages 91 - 92) have remainders. Students solve word problems by reading a problem, drawing a solution, and writing the appropriate division number sentence. For the *Picture This: Division* problems, they also write the multiplication number sentence. The directions below assume you will print the activity sheets for each student. To save paper, you can display them for the class and ask the students to draw their pictures on dry-erase boards or in math journals. You can download additional problems at **www.lauracandler.com/mmf** and create more problems using the templates (pages 93 - 94). **ONLINE**

Materials ✏️ ✂️ 🧵

- *Picture This: Division* and/ or *Picture This: Division Challenge* activity sheets for each student

Approximate Time:
5 minutes per problem

Procedure

1 Seat students in pairs and give all of your students the same *Picture This: Division* problem.

2 Have students read the problem and draw the solution individually. Remind them to keep their drawings simple and include only the details needed to solve the problem. Check to be sure they have done this before moving on.

3 Ask partners to compare and discuss their drawings with each other. If students find errors in their work, they may correct them.

4 Ask students to write the division number sentence, including the solution, to represent the problem. The *Picture This: Division* problems ask for the related multiplication number sentence, but the *Challenge*

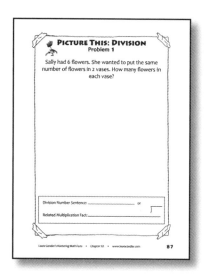

problems do not. When they are finished, have them compare and discuss with their partners.

5 Repeat steps 1 through 4 with the remaining problems.

6 When you are ready to introduce word problems with remainders, use the Division Challenge problems (pages 91 - 92) and follow steps 1 through 4.

LAURA'S Tips

After having students complete the first problem with a partner, you may want them to complete the remaining problems on their own. Use the blank templates (pages 93 - 94) to create additional word problems that have different levels of difficulty.

PICTURE THIS: DIVISION
Problem 1

Sally had 6 flowers. She wanted to put the same number of flowers in 2 vases. How many flowers in each vase?

Division Number Sentence: _____ **OR** ⌐‾‾‾‾

Related Multiplication Fact: _____

PICTURE THIS: DIVISION
Problem 2

Randy had 12 apples. He put an equal number of apples into 3 different baskets. How many apples were in each basket?

Division Number Sentence: _____ **OR** ⌐_____

Related Multiplication Fact: _____

PICTURE THIS: DIVISION
Problem 3

Jamal had 20 pieces of candy. He gave the same amount of candy to each of his 4 friends. How many pieces did he give to each friend?

Division Number Sentence: _____ OR ⌐_____

Related Multiplication Fact: _____

PICTURE THIS: DIVISION
Problem 4

Tameka bought some boxes of crayons. There were 24 crayons in all and each box had 8 crayons. How many boxes did she buy?

Division Number Sentence: _____ **OR** $\big)$‾

Related Multiplication Fact: _____

PICTURE THIS: DIVISION CHALLENGE
Problem 5

Ramon found 14 shells. He put an equal number in 3 different pails and had some left over. How many in each pail? How many left over?

Write it! ⟌

Krista baked 19 cookies. She put the same number on 4 plates and had a few left over. How many cookies on each plate? How many left over?

Write it! $\overline{)}$

PICTURE THIS: DIVISION

Division Number Sentence: _____ **OR** $\overline{\smash)}$

Related Multiplication Fact: _____

PICTURE THIS: DIVISION CHALLENGE

Write it!

CHAPTER 2

• • • • • • • • •

The Mastering Math
Facts System

The Mastering Math Facts System

• • • • • • • • •

Importance of Math Facts Fluency

By now your students recognize the importance of learning the times tables, and they are well on their way to developing a thorough understanding of what multiplication means. However, in order to solve advanced math problems quickly and accurately, students need to learn the times tables **fluently**. If they have to work out every basic math fact with arrays or skip counting, they will never become successful with more advanced math concepts.

The Mastering Math Facts system operates on the assumption that your students understand multiplication and division conceptually and they can apply those concepts to word problems. The system focuses on **developing computational fluency** through a variety of methods to practice times tables, track progress, and motivate students to master the math facts.

A key component of the Mastering Math Facts system is a motivational program to encourage and reward your students for learning the times tables fluently. I have developed two effective motivational programs, and you can use either one of them or modify them to create your own. Any motivational program you use should

> **Common Core State Standard for Multiplication and Division Fluency (CCSS.Math.Content.3.OA.C.7)**
>
> Fluently multiply and divide within 100, using strategies such as the relationship between multiplication and division (e.g., knowing that 8 × 5 = 40, one knows 40 ÷ 5 = 8) or properties of operations. By the end of Grade 3, know from memory all products of two one-digit numbers.

energize your students so that they'll be excited about mastering the math facts. Their feelings of accomplishment for learning the times tables will last long after they've enjoyed the reward they earned in your class!

The Mastering Math Facts system is proven to result in all of your students learning the times tables if the system is implemented according to the recommended guidelines and used consistently for 10 to 15 minutes a day. You are free to modify the components of the program to fit your own needs, but you should be aware that modifying the essential elements too much can give you less than optimal results.

Mastering Math Facts System Step-by-Step

1 DEFINE MATH FACTS MASTERY

2 EVALUATE READINESS AND FLUENCY

3 CHOOSE A MOTIVATIONAL PROGRAM

4 INTRODUCE THE MOTIVATIONAL PROGRAM

5 ASSESS MATH FACTS INDIVIDUALLY

6 TRACK PROGRESS

7 PRACTICE MIXED MATH FACTS DAILY

8 MONITOR STUDENT PROGRESS

9 RECOGNIZE INDIVIDUAL ACHIEVEMENT

10 CELEBRATE CLASS SUCCESS

STEP 1 — Define Math Facts Mastery

Before you start the system, you'll need to decide what it means for your students to demonstrate math fact mastery. This will depend on the grade level you teach, your mandated curriculum, and your students' proficiency when they arrive in your classroom. Some teachers require their students to master all times tables from 0 to 12, while others only from 0 to 9. You'll need to choose a time limit for the multiplication test your students will be required to complete with 100% accuracy. Use these suggested time limits for the Multiplication Facts Tests provided in Chapter 3 (pages 162 - 167), or modify them as needed:

0 TO 5 MULTIPLICATION FACTS TEST
 3rd grade - 4 minutes
 4th grade - 3 minutes
 5th grade - 2 minutes

0 TO 9 MULTIPLICATION FACTS TEST
 3rd grade - 5 minutes
 4th grade - 4 minutes
 5th grade - 3 minutes

0 TO 12 MULTIPLICATION FACTS TEST
 3rd grade - 6 minutes
 4th grade - 5 minutes
 5th grade - 4 minutes

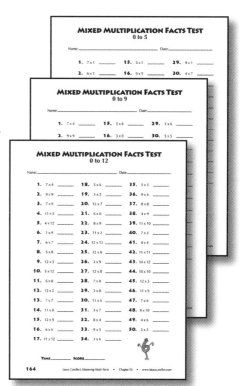

LAURA'S Tips

Learning the 12 times tables is helpful for measurement conversion involving feet and inches, as well as working with items sold in dozens. If you only require your students to learn the times tables up to 9, you can still encourage early finishers to continue practicing until they master all facts up to 12. Recognize students who make this extra effort with a special certificate during your class celebration.

STEP 2 — Evaluate Readiness and Fluency

Before you begin to implement the Mastering Math Facts system, you'll need to evaluate your students in two areas: readiness and fluency. First, you must determine their level of readiness. Starting students on this program before they are ready will result in frustration for you and your students.

Readiness means the level of understanding of the concepts of multiplication and division. You can assess your students' readiness by administering the informal assessment *Multiplication: Show What You Know* (page 11). If students do poorly on the assessment, spend more time on the activities in Chapter 1 and administer the *Multiplication: Show What You Know 2* (page 13) before beginning the Mastering Math Facts system.

Fluency refers to the speed and accuracy of students' ability to solve multiplication and division problems. You can assess fluency by administering a *Multiplication Facts Test* (page 164) to establish a baseline score for each student. Directions for how to administer such a timed math test are on page 102. The baseline score for fluency will include both time and number or percentage correct. The results of the fluency test will determine how to best implement the Mastering Math Facts system. Keep your students' scored Multiplication Facts Tests on file to refer to later, to determine if they are making adequate progress.

	What It Means	**How to Assess**	**Recommendations**
Readiness	Prior understanding of what multiplication and division mean on a conceptual level	Administer the informal assessment, *Multiplication: Show What You Know* (page 11)	If students do poorly on the assessment, spend time on the activities in Part 1 and administer the *Multiplication: Show What You Know Retest* before beginning the Mastering Math Facts system.
Fluency	How quickly and accurately students solve basic math problems, with particular emphasis on multiplication.	Administer one of the *Multiplication Facts Tests* (page 164) to establish a baseline score for each student. See page 102 for directions.	Use the results for the time and number correct on each student's fluency test to determine how to best implement the Mastering Math Facts system. Keep your students' scored Math Facts Tests on file to use as baseline assessment. You can refer to them later to determine if they are making adequate progress.

LAURA'S Tips

You can customize the Mastering Math Facts system by creating your own Multiplication Facts tests. You can create your own test sheets on the website **www.worksheetworks.com**, by entering the parameters for the problems and generating a worksheet with an answer key. You'll need to establish your own time limits for custom worksheets.

After you've evaluated your students' readiness and fluency, it's time to develop a plan for ensuring that all of your students know the math facts. If only a few students need additional practice, you can move on to Chapter 3 or 4 and provide your students with targeted multiplication practice activities or offer multiplication games in math centers. However, if they are just learning the math facts or a large number of your students are not proficient with math facts, you'll need to set aside 10 to 15 minutes a day to implement the complete Mastering Math Facts system.

Administer a Timed Math Facts Test

Procedure

1 Choose which Math Facts Test to give and duplicate one copy per student.

2 Distribute the papers and ask your students to keep their papers face down on their desks until it's time to begin.

3 Ask everyone to flip over their papers and begin writing the answers to the problems. Tell them not to write an equal sign, and to write the answers in order without skipping around. If they do get stuck they may skip a problem and come back to it later, but in general they need to do the problems in order to show real fluency.

4 As soon as they begin working, start the timer or stopwatch. Be sure it's set at 0 and is counting up in minutes and seconds.

5 Ask students to turn their papers face down and raise their hands when they finish. Remind them not to make distracting noises that will bother others who are still working.

Materials ✏️ ✂️ 📎

- Count up timer or stopwatch (preferably digital and displayed where students can see it)
- Math Facts Tests and Answer Keys (pages 162 - 167)

MIXED MULTIPLICATION FACTS TEST
0 to 9

Name:_____ Date:_____

1. 7 x 4 ____	15. 5 x 6 ____	29. 3 x 6 ____
2. 9 x 9 ____	16. 3 x 0 ____	30. 5 x 5 ____
3. 9 x 9 ____	17. 2 x 7 ____	31. 9 x 6 ____
4. 4 x 2 ____	18. 6 x 4 ____	32. 8 x 8 ____
5. 3 x 3 ____	19. 8 x 1 ____	33. 4 x 9 ____
6. 6 x 7 ____	20. 4 x 5 ____	34. 3 x 4 ____
7. 0 x 5 ____	21. 2 x 6 ____	35. 7 x 3 ____
8. 2 x 3 ____	22. 3 x 9 ____	36. 4 x 4 ____
9. 5 x 2 ____	23. 2 x 8 ____	37. 5 x 8 ____
10. 6 x 8 ____	24. 7 x 8 ____	38. 8 x 9 ____
11. 2 x 2 ____	25. 3 x 8 ____	39. 3 x 5 ____
12. 7 x 7 ____	26. 5 x 7 ____	40. 7 x 1 ____
13. 2 x 9 ____	27. 8 x 4 ____	
14. 6 x 6 ____	28. 9 x 5 ____	

TIME _____ **SCORE** _____

164   *Laura Candler's Mastering Math Facts* • Chapter XX • www.lauracandler.com

6 As each student's hand goes up, walk over to the students and record the stopping times on the paper. Alternatively, you could jot down their times on a class roster next to their names.

7 Allow time for most, if not all, students to finish. The first time you administer a timed fluency test, some students may not be able to complete it in a reasonable time. I recommend that you stop the testing period when 80% to 90% of your students are finished.

8 Score the tests using the appropriate Answer Key and record each student's percent correct or number correct next to his or her time.

STEP 3 Choose a Motivational Program

Choosing an effective motivational program is essential to the success of the Mastering Math Facts program. In fact, when you implement an effective motivational program systematically, you'll be amazed at how quickly your students will learn the math facts. I've found that there are a few essential components, and as long as you keep those elements in place, you can modify the motivational theme and add your own creative touches.

Essential components

- Individual and/or class system for assessing and tracking progress
- A whole class reward that all students are willing to work towards (special celebration, classroom party, field trip, etc.)
- A clear goal that must be met by each student to become a "Math Facts Master" (this can be individualized for special needs students)
- ALL students in the class must become a "Math Facts Master" for the class to earn the whole class reward

Motivational Theme Options

The Mastering Math Facts system includes complete directions and printables for two different motivational themes (pages 115-130). Choose a theme or adapt to create your own.

Theme 1: Here's the Scoop on Multiplication

- Whole class reward: Ice cream party when everyone has mastered all math facts
- Individual student goal: For each new math fact mastered, student adds one scoop of "ice cream" to his/her individual ice cream cone.

Theme 2: On Board to Multiplication Mastery

- Whole class reward: Field trip or special activity when everyone has mastered all math facts
- Individual student goal: For each new math fact mastered, student adds one "train car" to his/her individual train.

Modifications to Meet Individual Needs

As you plan the specific components of your motivational program, you'll need to consider how to differentiate instruction according to student needs. Here are a few suggestions:

ADVANCED LEARNERS ● Students who demonstrate mastery on the first fluency test should not be required to take part in the daily quizzes and math fact practices. Allow them to use this block of time to complete other assignments. As soon as they demonstrate fluency on a mixed practice test in the given time, let them add their completed ice cream cones or trains to the wall display. Give them their tickets to the party and present them with their Math Facts Master certificates right away. Doing so may motivate others to study harder and test out of the daily practice tests.

TIME LIMITS ● Students who work slowly or who are anxious about the time limits may be given extra time on the daily quizzes. For example, if you have a time limit of 1 minute on your math test, you might choose to give some students extra time or remove the time limit for those students.

ORAL RESPONSES ● Students who have difficulty with paper and pencil assignments might be allowed to respond orally or to use a computer program to demonstrate mastery.

INDIVIDUAL GOALS ● Set high expectations for all students, but be sure you have not made the class goal impossible for some students to reach. You can adapt the program by setting individual goals for each student and letting the class know that when every student has achieved his or her individual goal, the class will celebrate.

Procedure

1 SET UP THE CLASSROOM DISPLAY ● Before you introduce the program to your students, set up a classroom display to track the progress. Act mysterious about the program as you set up the display! This will create a sense of anticipation and excitement by not revealing all the details right away. You'll need a lot of wall space for the displays for either of these motivational programs, and they will look very colorful as the students fill them up. For the "Here's the Scoop on Multiplication" program, you'll start with a row of ice cream cones, each with a student's name. For the "On Board to Multiplication Mastery" program, you'll display a train engine for each student (see complete directions for creating these displays on page 115). Create a numbered key that shows the color that corresponds to each math fact.

2 DISCUSS IMPORTANCE OF MEMORIZING MATH FACTS ● When it's time to introduce your motivational program to your students, it's best to begin by being honest with them about why they need to learn the times tables fluently. Explain that as they get older and mathematics becomes more difficult, it's extremely important for them to know the times tables by heart. Share the analogy of learning to read: they will never become better readers if they have to sound out every word. In the same way, they will never be able to do advanced fractions and higher-level mathematics if they have to skip-count to find the math facts. Reassure them that you'll make the program as fun as possible with games, center activities, and online learning practice, but they must commit to doing whatever it takes to learn the times tables.

3 DESCRIBE YOUR "ALL OR NONE" MOTIVATIONAL PROGRAM ● Have your class display in place so that you can refer to it as you explain the program.

- Point to the wall display and explain that as your students learn each one of the math facts fluently, they will add an ice cream scoop to the cone with their name or a train car to their engine.

- Set the class goal of fluently learning the math facts to 9 or 12.

- Explain that when they learn all of the individual facts, they will be required to pass a mixed-practice test in a certain time limit. Reaching that goal makes a student a "Math Facts Master" who receives a ticket to the class celebration.

- Announce that when every student has mastered the required math facts, your class will celebrate with the whole-class goal (ice cream party, field trip, or special activity)!

4 REINFORCE WHOLE-CLASS GOAL • Many students are shocked when they hear that they won't get a class celebration unless everyone meets their individual goals. Invariably someone will ask, "What happens if we don't all learn the times tables?" My response? "Sorry - no party until you ALL learn them. We'll have to work together and help each other to make that happen, but I know you can do it." Stand firm on this one. You'll be amazed at how your students will work together in pursuit of this common class goal. Set a reasonable date for the celebration—generally a few months from when you begin the project. You may have to delay the party one or more times to ensure that all students meet the individual goal, so make sure you have several options for the whole-class goal noted on your calendar, even if you don't share those options with your students.

LAURA'S Tips

If you have special needs students in your classroom, you can set individual goals for students, rather than everyone having the same goal. If you choose to do this, let your students know that they have individual goals and when everyone meets his or her individual goal, the class will celebrate. In this case, you'll want to use a whole-class wall display as described on page 108.

STEP 5 Assess Math Facts Individually

In order for students to increase fluency with math facts, they have to practice them, and you have to assess their progress. You'll need to implement a method for practicing and assessing each math fact, one at a time.

Chapter 3 (page 139) includes complete descriptions of a number of methods for teachers to assess their students' progress toward learning the math facts. Methods include paper-and-pencil assessments, computer programs, and daily quizzes on individual dry-erase boards.

The daily assessment method that I've found most effective is the Daily Quick Quiz system (page 140). It's easy to implement and takes only minutes a day. Using this method, you can test your students quickly on one set of math facts each day. With the Daily Quick Quiz Method, students complete the work on individual dry-erase boards or on laminated, reusable pages. No paper needed!

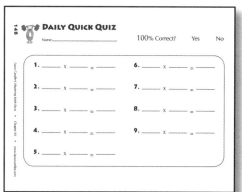

Be sure to choose a method that can be completed in just 10 to 15 minutes a day. Your math fact practice time could be at the beginning of every math class, or you might find 15 minutes later in the day, such as right before lunch or right after recess. In any event, it's best to establish a regular time for practice and stick to it until everyone has mastered the math facts. Whatever method you choose, it must provide the opportunity for students to practice a single set of math facts and take a quiz on one set of facts each day.

LAURA'S Tips

The major focus of the Mastering Math Facts system is on multiplication fact mastery. Students who have mastered the times tables usually master the division facts with ease.

STEP 6 Track Progress

One of the most important elements of the Mastering Math Facts system is how you track student progress. I find three different methods to be helpful: a one-page teacher chart, a wall display, and individual student records for each student. You can determine which methods best meet your needs.

1 TEACHER'S MASTER CHART ● This chart serves as your official record of your students' progress. Use a *Math Facts Record* (page 133 or 134) as your chart, and choose the version for your class—facts from 1 to 9 or from 2 to 12. Each day after your Daily Quick Quiz session, check off the math facts that your students have mastered. You can download slightly larger versions of these charts at **www.lauracandler.com/mmf**. ONLINE

2 WALL DISPLAYS ● In addition to keeping the *Math Facts Record* for your own records, create a classroom wall display that tracks all the students' progress in learning the math facts. Displaying a tracking system often serves to motivate and inspire students to do their best work. There are two different versions of wall displays that you can use:

INDIVIDUAL WALL DISPLAY ● Each student builds a separate ice cream cone or train on the wall where everyone can see it. Many teachers have reported that this kind of individual display encourages students to cheer each other on and help those who are struggling.

WHOLE CLASS WALL DISPLAY ● Some teachers prefer not to have displays that highlight individual progress. You can modify the display component by having one giant ice cream cone or train for

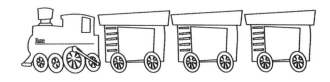

the whole class. Create a large cone or train engine, by enlarging the patterns on pages 117 and 125, to about the size of an 8-1/2 x 11-inch sheet of paper. Write the class name on the cone or train engine. Then add one large ice cream scoop or large train car when everyone in the class has mastered that particular math fact.

3 **INDIVIDUAL STUDENT RECORDS** ● Students also enjoy tracking their progress through their own private recording systems. Pages 120 and 128 are charts for this purpose, using the ice cream or the train theme. There are variations for each theme for mastering facts from 1 to 9 or 1 to 12. Each student stores the individual chart in a folder. Each day after you complete your Daily Quick Quiz and check off who has mastered which math facts, have your students get out their charts and color in the ice cream scoop or train that represents the math fact they mastered that day.

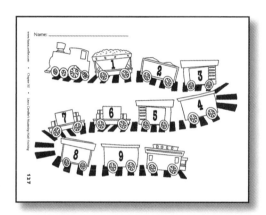

STEP 7 — Practice Mixed Math Facts Daily

It's one thing for students to memorize each set of math facts and write them fluently in a minute or less. But it's quite a different matter to demonstrate fluency when those math facts are mixed together! It happens time and again that a student will successfully build a tall ice cream cone or long train and then be unable to pass a practice test that mixes all the learned math facts together. To prevent this from happening, provide opportunities for students to practice mixed math facts on a regular basis.

Chapter 4 (page 173) provides a number of strategies that work well, including:

FLASH CARD PRACTICE (PAGE 174) ● Students can study flash cards on their own or paired with a partner (see directions on pages 175 - 176).

MOVEMENT (PAGE 174) ● Many students are kinetic learners, and math facts are a perfect vehicle for matching movement to practice activities!

TIMES TABLE CHALLENGES (PAGES 177 - 182) ● Completing variations of the standard times table charts are effective ways to challenge students to practice all the math facts.

TECHNOLOGY (PAGE 183) ● Free online websites, subscription software, and mobile device apps are all great ways to incorporate technology in practicing math facts.

MATH GAMES (PAGE 184) ● Encourage students to play math facts review games after they completed other assignments or in math centers.

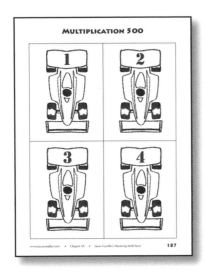

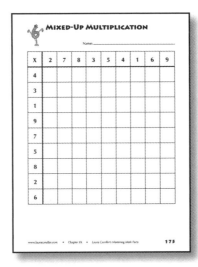

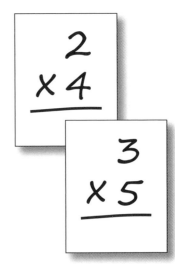

STEP 8 Monitor Student Progress

After your students have been working on the math facts for a week or so, look for patterns in student progress. If you have a wall display, it's easy to see these patterns, but you should also look at your own records to see who is making good progress and who is lagging behind. Some students dive into the process and systematically study each set of times tables until they've learned them fluently. They understand the importance of memorizing math facts and they master the multiplication tables in just a few weeks. Other students will not progress as quickly. You'll need a variety of strategies to deal with different rates of student progress:

ENCOURAGE RESISTANT STUDENTS ● Some students are very resistant to learning the times tables, especially if they have tried to learn them before and were not successful. These students need a lot of encouragement and motivation to ensure that they don't give up. You can ask other students to play review games with them in their free time. If you have access to tablets or other mobile devices, let them practice the math facts with a fun review app.

INTERVENE EARLY ● Don't let anyone fall too far behind the others without intervening. Pull struggling students aside for some extra help or ask a parent volunteer to work with them to practice with flash cards. Some students become stressed when asked to take a timed test. If these students know the math facts accurately but are anxious about the time limit, give them a little extra time on the daily quizzes.

CONTACT PARENTS ● Be sure to contact parents as soon as you notice any of your students struggling to master the math facts. Send a letter (page 131) and a *Multiplication Log* (page 132) home with those students. Require them to complete one section of the log each day until they have mastered all the facts. To make this homework more fun and engaging, prepare some of the games in Chapter 4 as activities for students to check out and take home to play with their families.

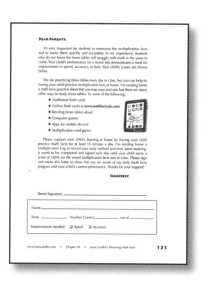

BE CONFIDENT! ● Above all, let your struggling students know that you won't give up on them and you believe that they can learn the math facts. When you express confidence in them and show that are willing to help, they will begin to believe in themselves, too.

# STEP 9 Recognize Individual Achievement

As your students see their ice cream cones grow taller or their trains grow longer, they will feel a sense of accomplishment. This is the time to encourage them to complete the process and become a Math Facts Master. To do so, your students must complete two steps: learn all the math facts individually and pass the Mixed Math Facts test.

The first step to learning all the facts individually is completed when all nine or twelve scoops of ice cream or train cars are on a student's display. To become a Math Facts Master, students must score 100% on the Mixed Math Facts Test (pages 162 - 167). Administer it according to the directions on page 102. If students don't score 100% in the allotted time, give them a chance to review incorrect answers and allow them to try again the next day.

After students successfully pass the Mixed Math Facts Test, it's important to recognize them individually as Math Facts Masters (the whole class celebration will come when *everyone* has passed the test). There are several options for recognizing a student when he/she becomes an individual Math Facts Master:

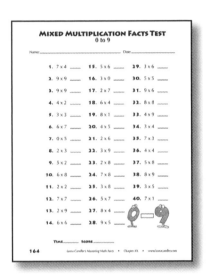

- Place a blob of "whipped cream" on the top ice cream scoop or a "caboose" on the train

- Give the student a ticket to the class celebration (pages 122 and 130)

- Present the student with a Math Facts Master award certificate (page 135). You can download a customizable certificate at **www.lauracandler.com/mmf**.
  ⌂ONLINE

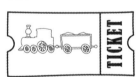

STEP 10 Celebrate Class Success

When everyone in the class has passed the Math Facts Mastery Test (or reached their individual goals), celebrate! An ice cream party is the celebration for the "Here's the Scoop on Multiplication" method, and a field trip or special activity works well for the "On Board to Multiplication Mastery" method. The key to success is to announce the celebration at the beginning of your motivational program and to be sure to follow through when everyone has completed the program. This is a time to celebrate a job well done!

LAURA'S Tips

What do you do if you have a student who does not reach the individual goal long after the rest of the class has met theirs? Should you have the celebration and leave that student out? NO! I've used this program for more than ten years and I've never had a student who refused to cooperate or was unable to master the required math facts. I did have to modify my expectations for students with special needs, but they all reached their individual goals. But if I ever had this situation occur, I would not hold the whole class celebration. Instead, I would find another way to reward all the students who met their individual goals. Students need to know that when the teacher makes a promise, he/she will follow through, even when it's hard to do.

Mastering Math Facts System
Directions and Printables

"Here's the Scoop on Multiplication" Display Directions

Individual Student Wall Display

PREPARATION AND ASSEMBLY:

1 Using the patterns on page 117, cut out one beige ice cream cone for each student and write his or her name at the top.

2 Decide whether you are going to require students to learn math facts to 9 or 12. For each math fact, choose a different color of card stock.

3 Using the pattern on page 118, duplicate enough ice cream scoops so that each student will have one of every color.

4 Duplicate one blob of whipped cream for each student.

5 Laminate and cut out the scoops and whipped cream blobs and store them in labeled plastic zipper bags. Label each bag with the math fact that color represents. (Example: 1s are white, 2s are red, etc.)

Materials ✏ ✂ ✂

- Card stock in a variety of colors
- Ice cream patterns (pages 117 - 119)

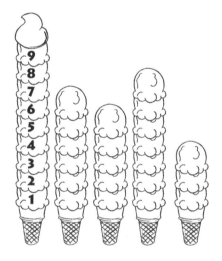

6 Place the cones with student names along the bottom of your display area.

7 Make a key down one side of the display that shows the color that represents each math fact.

8 As each student masters a math fact, he or she adds a scoop to the top of his or her ice cream cone.

9 When each student passes the timed mixed practice test, the student adds a blob of whipped cream on the top of his/her ice cream cone.

Whole Class Wall Display

PREPARATION AND ASSEMBLY:

1 Enlarge the ice cream cone pattern (page 117) to the size of an 8-1/2 x 11-inch sheet of paper. Trace or duplicate the pattern on card stock and cut it out. Label it with the class name. Large-sized patterns can be downloaded at **www.lauracandler.com/mmf.**

ONLINE

2 Decide whether you are going to require students to learn math facts to 9 or 12. For each math fact, choose a different color of card stock.

3 Enlarge the ice cream scoop pattern (page 118) and create one large scoop in every color. Label each scoop with the number of the math fact it represents.

4 Enlarge the whipped cream pattern (page 119) to create one large blob of whipped cream.

5 Laminate the pieces so that they can be used year after year.

6 Place the ice cream cone with class name near the bottom of your vertical display area.

7 When everyone in the class masters one of the math facts, add that scoop to the top of the class ice cream cone.

8 When all math facts are mastered and all students have passed the timed mixed practice test, add the blob of whipped cream to the top scoop.

LAURA'S Tips

I don't recommend using construction paper because it quickly fades and looks outdated after one year. To find enough colors for your display, you'll probably need to use both bright and pastel colors of card stock.

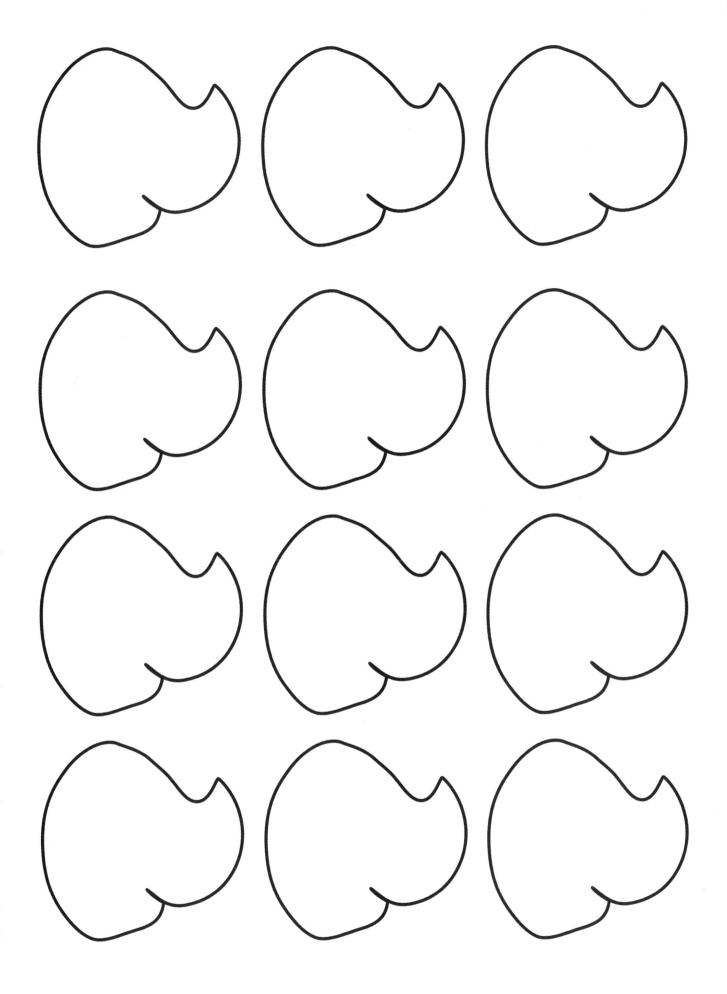

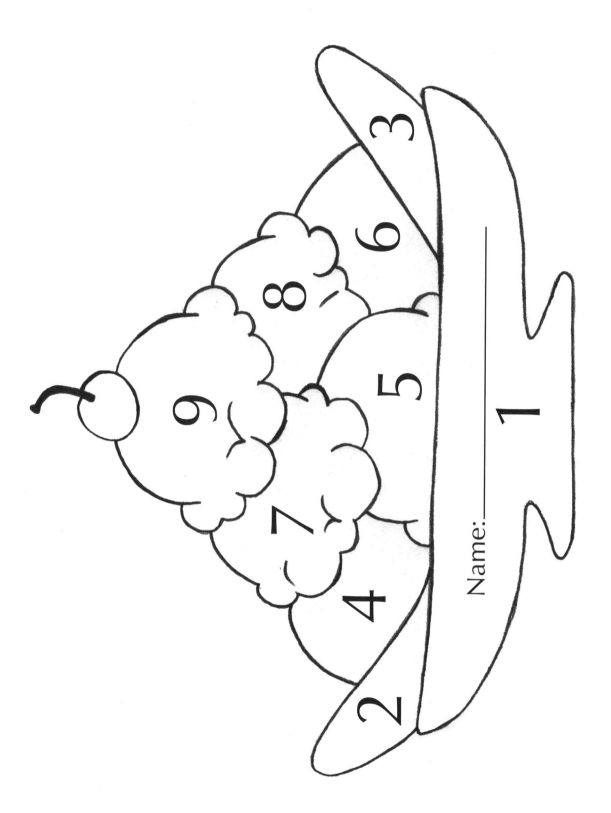

Name: _____

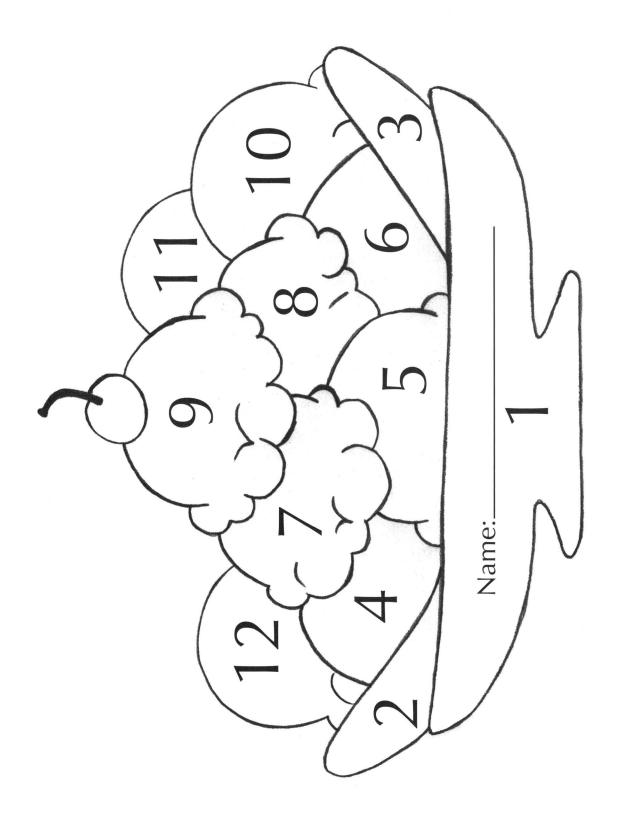

Name: _____

"On Board to Multiplication Mastery" Display Directions

LAURA'S Tips

I don't recommend using construction paper because it quickly fades and looks outdated after one year. To find enough colors for your display, you'll probably need to use both bright and pastel colors of card stock.

Materials ✐ ✄ ✎

- Card stock in a variety of colors
- Train patterns (pages 125 - 127)

Individual Student Wall Display

PREPARATION AND ASSEMBLY:

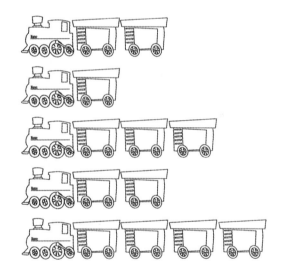

1 Using the patterns on page 125, cut out one train engine for each student and write his or her name on it.

2 Decide whether you are going to require students to learn math facts to 9 or 12. For each math fact, choose a different color or shade of card stock.

3 Using the pattern on page 126, duplicate and laminate enough train cars so that each student will have one of every color.

4 Duplicate and laminate one caboose (page 127) for each student.

5 Cut out the train cards and cabooses and store them in labeled plastic zipper bags. Label each bag with the math fact that color represents. (Example: 1s are white, 2s are red, etc.)

6 Place the train engines with student names along the left side of your display area.

7 Make a key near your display that shows the color that represents each math fact.

8 As each student masters a math fact, he/she adds a train car to the end of his or her individual train.

9 When each student passes the timed mixed practice test, the student adds the caboose to the end.

Whole Class Wall Display

PREPARATION AND ASSEMBLY:

1 Enlarge the train engine pattern on page 125 to the size of an 8-1/2 x 11-inch sheet of paper. Duplicate it on card stock and label it with the class name. Large-sized patterns can be downloaded at **www.lauracandler.com/mmf.**
ONLINE

2 Decide whether you are going to require students to learn math facts to 9 or 12. For each math fact, choose a different color or shade of card stock.

3 Enlarge the train car pattern (page 126) and create one car in every color. Label each car with the number of the math fact it represents.

4 Enlarge the caboose pattern (page 127) to create one large caboose.

5 Laminate the pieces so that they can be used year after year.

6 Place the engine with class name on the left side of your horizontal display area.

7 As everyone in the class masters one of the math facts, add that train car to the end of the class train.

8 When all math facts are mastered and all students have passed the timed mixed practice test has been passed, add the caboose to the end.

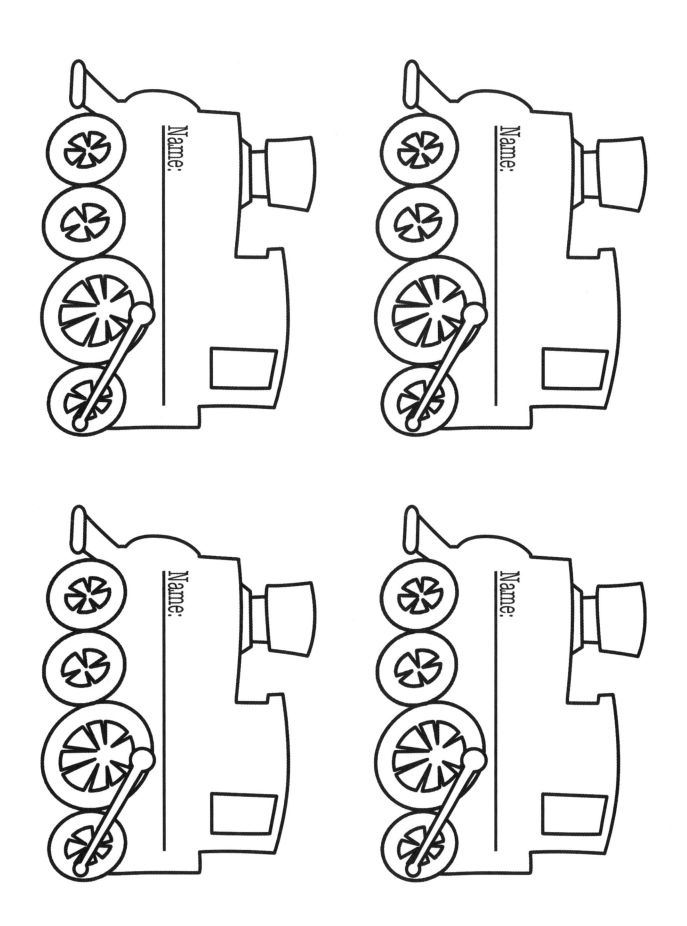

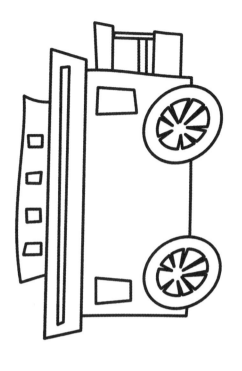
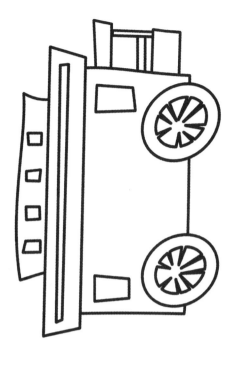
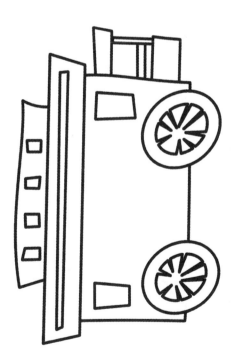
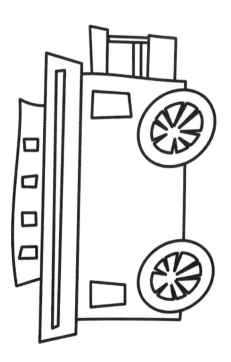

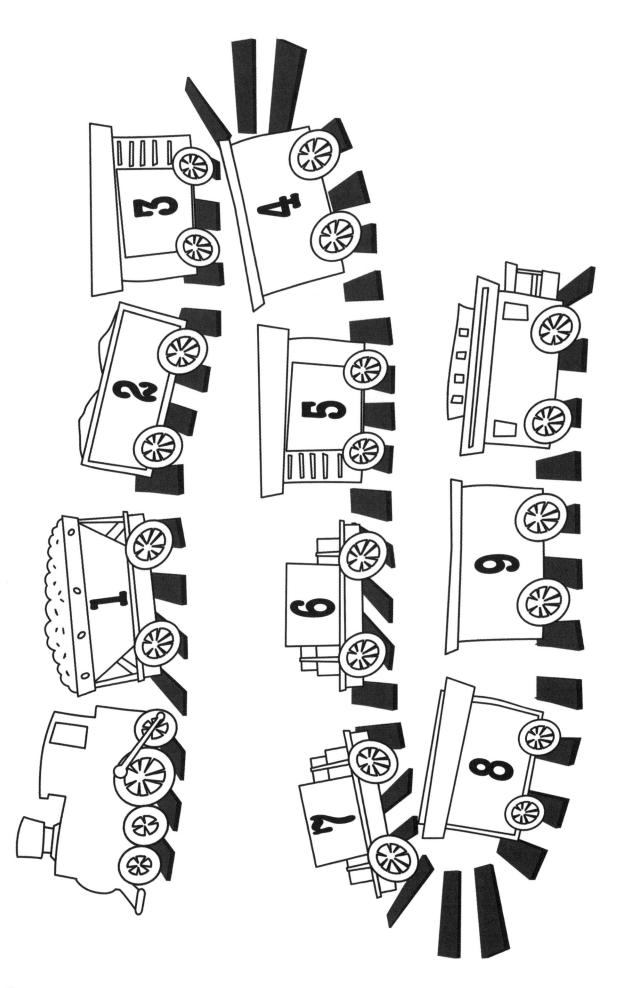

Name: _____

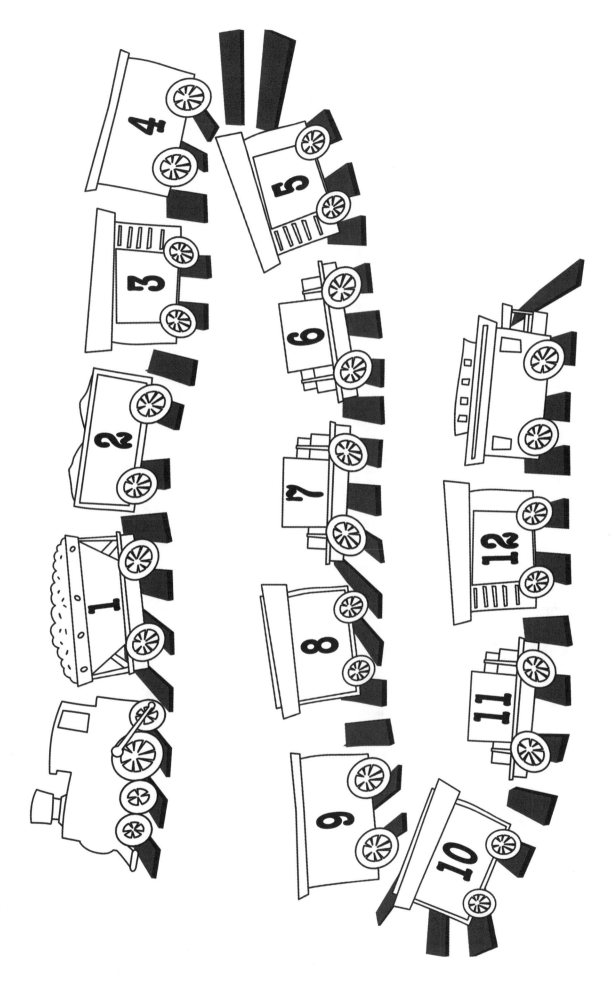

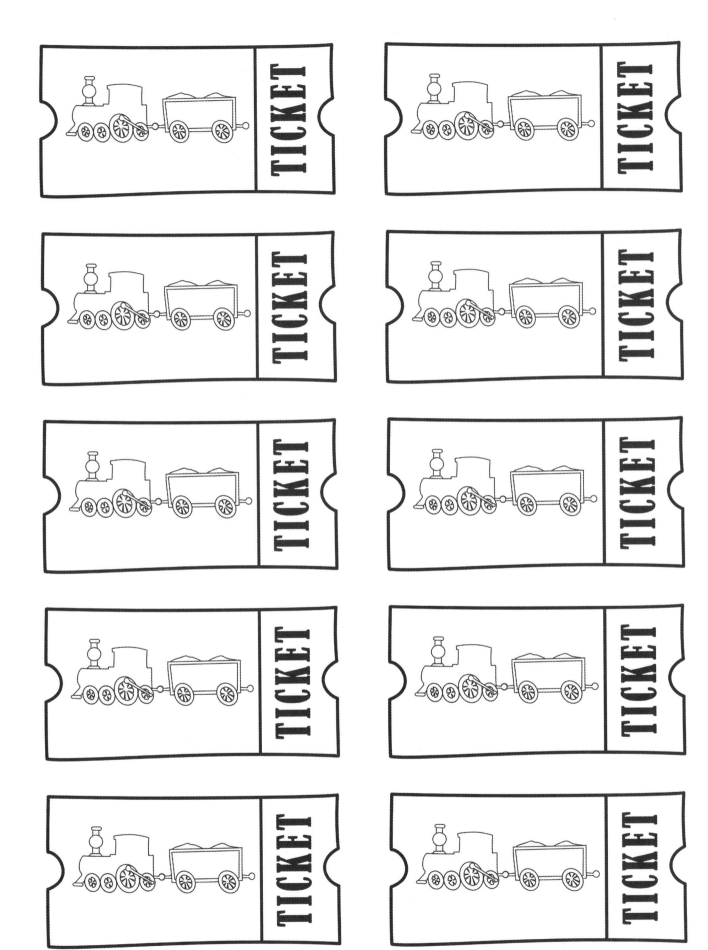

DEAR PARENTS,

It's very important for students to memorize the multiplication facts and to know them quickly and accurately. In my experience, students who do not know the times tables will struggle with math in the years to come. Your child's performance on a recent test demonstrates a need for improvement in speed, accuracy, or both. Your child's scores are shown below.

We are practicing times tables every day in class, but you can help by having your child practice multiplication facts at home. I'm sending home a math facts practice sheet that you may copy and use, but there are many other ways to study times tables. Try some of the following:

- Traditional flash cards
- Online flash cards at **www.mathfactcafe.com**
- Reciting times tables aloud
- Computer games
- Apps for mobile devices
- Multiplication card games

Please support your child's learning at home by having your child practice math facts for at least 15 minutes a day. I'm sending home a Multiplication Log to record your study method and time spent studying. It needs to be completed and signed each day until your child earns a score of 100% on the timed multiplication facts test in class. Please sign and return this letter to show that you are aware of my daily math facts program and your child's current performance. Thanks for your support!

SINCERELY,

Parent Signature: _____

Name:_____

Time: _____ Number Correct:_____ out of _____

Improvements needed: ❏ Speed   ❏ Accuracy

MULTIPLICATION
Study Log

Name: _____

Week of: _____

Mastering multiplication facts is essential at this grade level! In order to learn to divide and reduce fractions, your child needs to know the times tables quickly and without having to count on fingers. Please study with your child at least 15 minutes a day until these facts are mastered. Keep it fun by using different methods each day. Try flash cards, math practice sheets or workbooks, oral review (calling out problems and answers), and websites like **www.mathfactcafe.com**.

WEEKEND	Study Method: _____ _____ _____ _____	Study Minutes Parent Initials
MONDAY	Study Method: _____ _____ _____ _____	Study Minutes Parent Initials
TUESDAY	Study Method: _____ _____ _____ _____	Study Minutes Parent Initials
WEDNESDAY	Study Method: _____ _____ _____ _____	Study Minutes Parent Initials
THURSDAY	Study Method: _____ _____ _____ _____	Study Minutes Parent Initials

MATH FACTS RECORD 1 - 9	1	2	3	4	5	6	7	8	9
1.									
2.									
3.									
4.									
5.									
6.									
7.									
8.									
9.									
10.									
11.									
12.									
13.									
14.									
15.									
16.									
17.									
18.									
19.									
20.									
21.									
22.									
23.									
24.									
25.									

Math Facts Record 2 - 12	2	3	4	5	6	7	8	9	10	11	12
1.											
2.											
3.											
4.											
5.											
6.											
7.											
8.											
9.											
10.											
11.											
12.											
13.											
14.											
15.											
16.											
17.											
18.											
19.											
20.											
21.											
22.											
23.											
24.											
25.											

Certificate of Award

CONGRATULATIONS TO

for becoming a

MATH FACTS MASTER

With hard work and study you mastered the times tables and earned this special recognition!

_____ Signature

_____ Date

CHAPTER 3

• • • • • • • • •

Math Facts
Assessment Strategies

Math Facts Assessment Strategies

● ● ● ● ● ● ● ● ●

The Mastering Math Facts system includes a key element: a way for teachers to assess math fact fluency. There are two different methods for assessing students' progress as they learn the individual facts—Daily Quick Quizzes and Mini Math Checks. Choose either method that works for you; both have advantages. The Daily Quick Quiz saves paper if you have individual dry-erase boards available in your classroom. Mini Math Checks are easier to implement, but they do require quite a bit of paper.

No matter which method you use, you'll need to keep a progress chart using the Math Facts Record (page 133 or 134) and check off math facts as students master them.

After students have mastered the math facts one at a time, use the third type of assessment—mixed math facts tests (pages 162 - 167)—to assess students' fluency and accuracy when all the facts are mixed together. Using these assessments will ensure that your students are meeting Common Core Content Standard 3.OA.C.7: Fluently multiply and divide within 100.

- **DAILY QUICK QUIZ (PAGES 140 - 152)**

- **MINI MATH CHECK (PAGES 153 - 159)**

- **MIXED MATH FACTS ASSESSMENTS (PAGES 160 - 169)**

Individual Math Facts Assessment 1
Daily Quick Quiz

The Daily Quick Quiz is a simple and effective way to assess students' knowledge of the individual math facts. This method can be implemented in as little as ten minutes a day. To save paper, your students can use individual dry-erase boards and markers to record their math problems and answers. If you don't have individual dry-erase boards, you can cut scrap paper into ¼-sheets. Provide stacks of these papers so each student can easily get a sheet every day.

The Daily Quick Quiz has several steps that must be followed in order, but after you administer a few quizzes, the process will be easy. The first time you give a Daily Quick Quiz, display the steps (pages 143 - 147) one at a time to your class as you explain the procedure. Each step gives the directions and shows what the students' dry-erase boards will look like when that step is completed. The steps illustrate how to assess math facts from 1 to 9, but if your class is learning facts 1 to 12, have students number their boards from 1 to 12.

Procedure

1 **PREPARE BOARD** ● Students number their dry-erase boards or scrap paper from 1 to 9 or 12 as shown. Students write their names at the bottom.

2 **LIST MATH FACT PROBLEMS** ● For each number, students write the math fact to be tested (7 x ____ =   ).

Materials ✏️ ✂️ 🧷

- Dry-erase board and marker (or ¼-sheet scrap paper) for each student
- Countdown timer
- Random factors list (pages 148 - 149)

STEP 1

1.	6.
2.	7.
3.	8.
4.	9.
5.	*Cindy*

STEP 2

1. 7 x ____ =	6. 7 x ____ =
2. 7 x ____ =	7. 7 x ____ =
3. 7 x ____ =	8. 7 x ____ =
4. 7 x ____ =	9. 7 x ____ =
5. 7 x ____ =	*Cindy*

3 ADD THE MISSING FACTORS ●

Call out the random factors from 1 to 9 that go in each blank. Use a new list of random factors (pages 148 - 149) each day so you call them out in a different order every time. As you call out the factors, students record them in that order on the blank lines in each problem. Ask them to turn their dry-erase boards or papers face down when ready. Step 3 shows a board ready for the quiz.

4 TIME STUDENTS ●

When everyone is ready with their boards face down, call "Start!" Students flip their boards or papers face up and begin filling in the blanks in order without skipping problems. They must complete all answers in 1 minute or whatever time you have previously announced. Tell them to turn their boards or papers face down when they finish. Step 4 shows a completed board or papers at the end of the quiz.

5 TRADE AND CHECK ●

When time is up, ask students who have finished to trade boards or papers and check each other's answers as shown. Students who did not finish on time may check their own answers and correct any that they missed. Students can check using one of these methods:

- Listen and check as teacher reads answers aloud
- Use a calculator
- Refer to times table chart
- Use an answer key

After students practice checking each other's work a few times, they should be able to do it quickly and accurately. If you prefer not to have students check each other's work, have them stack their boards or papers on a table for you or an assistant to check.

STEP 3

1. $7 \times \underline{2} =$ 6. $7 \times \underline{9} =$

2. $7 \times \underline{7} =$ 7. $7 \times \underline{3} =$

3. $7 \times \underline{8} =$ 8. $7 \times \underline{6} =$

4. $7 \times \underline{5} =$ 9. $7 \times \underline{4} =$

                          Cindy

5. $7 \times \underline{1} =$

STEP 4

1. $7 \times \underline{2} = 14$ 6. $7 \times \underline{9} = 63$

2. $7 \times \underline{7} = 49$ 7. $7 \times \underline{3} = 21$

3. $7 \times \underline{8} = 56$ 8. $7 \times \underline{6} = 42$

4. $7 \times \underline{5} = 35$ 9. $7 \times \underline{4} = 28$

                               Cindy

5. $7 \times \underline{1} = 7$

STEP 5

1. $7 \times \underline{2} = 14$ ✓   6. $7 \times \underline{9} = 63$ ✓

2. $7 \times \underline{7} = 49$ ✓   7. $7 \times \underline{3} = 21$ ✓

3. $7 \times \underline{8} = 56$ ✓   8. $7 \times \underline{6} = 42$ ✓

4. $7 \times \underline{5} = 35$ ✓   9. $7 \times \underline{4} = 28$ ✓

                               Cindy

5. $7 \times \underline{1} = 7$ ✓   *Checked by Xavier*

6 **RECORD PROGRESS** ● Ask any student who scores a 100% to bring their completed board or paper to you so you can record the score. Use a Math Facts Record (page 133 or 134) and place a check in the column under the math fact they have mastered.

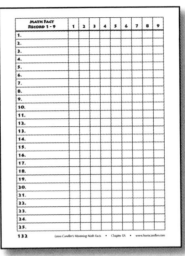

LAURA'S Tips

- Daily Quick Quizzes may be given at the same time to students who are learning different math facts.

- Students who have mastered a given set of math facts can be designated as experts for that math fact. Experts may assist other students by helping to check their answers for the math facts they know well.

- To save set-up time, print and laminate the blank Daily Quick Quiz template (page 150 or 151) on white card stock, one for each student. Students can write on these with dry-erase markers. (Thanks to third grade teacher Suzy Brooks of Falmouth, Massachusetts for this great tip!)

  OR:

  Print lots of copies of the Daily Quick Quiz strips (page 152) and have those available for your students to grab when it's time to take the quiz.

STEP 1

Number your dry-erase board.
Write your name at the bottom.

1.

2.

3.

4.

5.

6.

7.

8.

9.

Cindy

STEP 2

List number sentences with blanks for the math fact you want to master.

1. 7 x _____ =

2. 7 x _____ =

3. 7 x _____ =

4. 7 x _____ =

5. 7 x _____ =

6. 7 x _____ =

7. 7 x _____ =

8. 7 x _____ =

9. 7 x _____ =

Cindy

STEP 3

Fill in the blanks with random factors from 1 to 9.
Turn your board face down to show you are ready.

1. 7 x ___2___ =

2. 7 x ___7___ =

3. 7 x ___8___ =

4. 7 x ___5___ =

5. 7 x ___1___ =

6. 7 x ___9___ =

7. 7 x ___3___ =

8. 7 x ___6___ =

9. 7 x ___4___ =

Cindy

STEP 4

Without skipping problems, write the products before the time runs out. Turn your board face down when you finish.

1. $7 \times \underline{2} = 14$ **6.** $7 \times \underline{9} = 63$

2. $7 \times \underline{7} = 49$ **7.** $7 \times \underline{3} = 21$

3. $7 \times \underline{8} = 56$ **8.** $7 \times \underline{6} = 42$

4. $7 \times \underline{5} = 35$ **9.** $7 \times \underline{4} = 28$

                                              Cindy

5. $7 \times \underline{1} = 7$

STEP 5

When time is up, trade boards and correct your partner's answers. Sign your name as the Checker.

1. 7 x ___2___ = 14 ✓   **6.** 7 x ___9___ = 63 ✓

2. 7 x ___7___ = 49 ✓   **7.** 7 x ___3___ = 21 ✓

3. 7 x ___8___ = 56 ✓   **8.** 7 x ___6___ = 42 ✓

4. 7 x ___5___ = 35 ✓   **9.** 7 x ___4___ = 28 ✓

Cindy

5. 7 x ___1___ = 7 ✓   *Checked by Xavier*

Random Factor Lists for Daily Quick Quizzes (1 to 9)

When administering a Daily Quick Quiz, call out a list of factors after students have prepared their dry-erase boards or papers as directed. Having a different list of factors for each day of the week will prevent students from memorizing answers in a certain order.

Monday	Tuesday	Wednesday	Thursday	Friday
4	2	5	3	9
1	4	2	2	5
3	7	9	1	2
7	6	7	9	7
6	9	6	4	1
9	3	8	7	6
2	5	4	6	3
5	1	3	8	8
8	8	1	5	4

Random Factor Lists for Daily Quick Quizzes (1 to 12)

When administering a Daily Quick Quiz, call out a list of factors after students have prepared their dry-erase boards or papers as directed. Having a different list of factors for each day of the week will prevent students from memorizing answers in a certain order.

Monday	Tuesday	Wednesday	Thursday	Friday
4	2	5	10	9
1	4	2	5	5
3	11	9	2	2
11	7	11	1	10
7	10	7	9	7
6	6	10	4	12
9	9	6	11	1
2	12	8	7	6
12	3	12	12	3
5	5	4	6	8
8	1	3	8	4
10	8	1	3	11

DAILY QUICK QUIZ

Name: _____

100% Correct?   Yes   No

1. _____ × _____ = _____

2. _____ × _____ = _____

3. _____ × _____ = _____

4. _____ × _____ = _____

5. _____ × _____ = _____

6. _____ × _____ = _____

7. _____ × _____ = _____

8. _____ × _____ = _____

9. _____ × _____ = _____

Daily Quick Quiz

Name: _____

100% Correct?   Yes   No

1. ___ × ___ = ___
2. ___ × ___ = ___
3. ___ × ___ = ___
4. ___ × ___ = ___
5. ___ × ___ = ___
6. ___ × ___ = ___
7. ___ × ___ = ___
8. ___ × ___ = ___
9. ___ × ___ = ___
10. ___ × ___ = ___
11. ___ × ___ = ___
12. ___ × ___ = ___

DAILY QUICK QUIZ

Name: _____

Multiplication Fact: _____

1. ___ × ___ = ___
2. ___ × ___ = ___
3. ___ × ___ = ___
4. ___ × ___ = ___
5. ___ × ___ = ___
6. ___ × ___ = ___
7. ___ × ___ = ___
8. ___ × ___ = ___
9. ___ × ___ = ___
10. ___ × ___ = ___
11. ___ × ___ = ___
12. ___ × ___ = ___

100% Correct?   Yes   No

DAILY QUICK QUIZ

Name: _____

Multiplication Fact: _____

1. ___ × ___ = ___
2. ___ × ___ = ___
3. ___ × ___ = ___
4. ___ × ___ = ___
5. ___ × ___ = ___
6. ___ × ___ = ___
7. ___ × ___ = ___
8. ___ × ___ = ___
9. ___ × ___ = ___
10. ___ × ___ = ___
11. ___ × ___ = ___
12. ___ × ___ = ___

100% Correct?   Yes   No

DAILY QUICK QUIZ

Name: _____

Multiplication Fact: _____

1. ___ × ___ = ___
2. ___ × ___ = ___
3. ___ × ___ = ___
4. ___ × ___ = ___
5. ___ × ___ = ___
6. ___ × ___ = ___
7. ___ × ___ = ___
8. ___ × ___ = ___
9. ___ × ___ = ___
10. ___ × ___ = ___
11. ___ × ___ = ___
12. ___ × ___ = ___

100% Correct?   Yes   No

Individual Math Facts Assessment 2
Mini Math Check

Each Mini Math Check (pages 154 - 156) assesses one set of math facts in random order. You can use a Mini Math Check to test the class all at once, or allow students to work with a partner in a math center to test each other.

Procedure

1 PREPARE MINI MATH CHECKS • Duplicate enough Mini Math Checks for your class, cut them apart, and store them in envelopes labeled with the math fact number.

2 ATTACH ANSWER KEY • To make the activity self-checking, glue an answer key on the back of the envelope.

3 SET TIME LIMIT • Give a time limit, such as one minute, for completing the sheet. Determine the time limit based on grade level and students' proficiency.

4 TIME STUDENTS • When everyone is ready, call "Start!" Students begin filling in the blanks in order without skipping problems. They must complete all answers in 1 minute or whatever time you have previously announced. Tell them to turn their papers face down when they finish.

5 TRADE AND CHECK • After students complete the Mini Math Check, grade them yourself or let the students correct each other's work using the answer key.

6 RECORD PROGRESS • Ask any student who scores a 100% to bring their completed paper to you so you can record the score. Use a Math Facts Record (page 133 or 134) and place a check in the column under the math fact they have mastered.

MINI MATH CHECKS

Name:_____

 2 x 4 = _____

2 x 3 = _____

2 x 6 = _____

2 x 5 = _____

2 x 8 = _____

2 x 7 = _____

2 x 2 = _____

2 x 9 = _____

Name:_____

3 x 3 = _____

3 x 9 = _____

3 x 5 = _____

3 x 2 = _____

3 x 4 = _____

3 x 6 = _____

3 x 8 = _____

3 x 7 = _____

Name:_____

 4 x 8 = _____

4 x 3 = _____

4 x 5 = _____

4 x 7 = _____

4 x 9 = _____

4 x 2 = _____

4 x 4 = _____

4 x 6 = _____

Name:_____

5 x 8 = _____

5 x 2 = _____

5 x 3 = _____

5 x 9 = _____

5 x 5 = _____

5 x 7 = _____

5 x 4 = _____

5 x 6 = _____

MINI MATH CHECKS

Name:_____

 6 x 6 = _____

 6 x 3 = _____

 6 x 7 = _____

 6 x 2 = _____

 6 x 4 = _____

 6 x 9 = _____

 6 x 8 = _____

 6 x 5 = _____

Name:_____

 7 x 5 = _____

 7 x 3 = _____

 7 x 9 = _____

 7 x 4 = _____

 7 x 2 = _____

 7 x 8 = _____

 7 x 6 = _____

 7 x 7 = _____

Name:_____

8 x 8 = _____

8 x 6 = _____

8 x 4 = _____

8 x 3 = _____

8 x 9 = _____

8 x 2 = _____

8 x 5 = _____

8 x 7 = _____

Name:_____

 9 x 2 = _____

 9 x 9 = _____

 9 x 7 = _____

 9 x 3 = _____

 9 x 5 = _____

 9 x 8 = _____

 9 x 4 = _____

9 x 6 = _____

MINI MATH CHECKS

Name:_____

Name:_____

 11 x 10 = _____

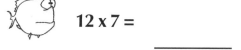 **12 x 10 =** _____

 11 x 6 = _____

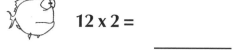 **12 x 6 =** _____

 11 x 3 = _____

12 x 3 = _____

 11 x 7 = _____

 **12 x 7 =** _____

 11 x 2 = _____

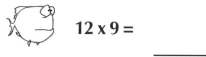 **12 x 2 =** _____

 11 x 12 = _____

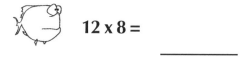 **12 x 12 =** _____

 11 x 9 = _____

12 x 9 = _____

 11 x 8 = _____

12 x 8 = _____

 11 x 5 = _____

12 x 5 = _____

 11 x 11 = _____

 12 x 11 = _____

 11 x 4 = _____

12 x 4 = _____

MINI MATH CHECKS

2s Answer Key

 2 x 4 = ___8___

2 x 3 = ___6___

2 x 6 = ___12___

2 x 5 = ___10___

2 x 8 = ___16___

 2 x 7 = ___14___

 2 x 2 = ___4___

 2 x 9 = ___18___

3s Answer Key

3 x 3 = ___9___

3 x 9 = ___27___

3 x 5 = ___15___

3 x 2 = ___6___

3 x 4 = ___12___

3 x 6 = ___18___

3 x 8 = ___24___

3 x 7 = ___21___

4s Answer Key

4 x 8 = ___32___

4 x 3 = ___12___

4 x 5 = ___20___

4 x 7 = ___28___

 4 x 9 = ___36___

 4 x 2 = ___8___

4 x 4 = ___16___

4 x 6 = ___24___

5s Answer Key

 5 x 8 = ___40___

 5 x 2 = ___10___

 5 x 3 = ___15___

 5 x 9 = ___45___

 5 x 5 = ___25___

 5 x 7 = ___35___

 5 x 4 = ___20___

5 x 6 = ___30___

MINI MATH CHECKS

6s Answer Key

 6 x 6 = **36**

 6 x 3 = **18**

 6 x 7 = **42**

 6 x 2 = **12**

 6 x 4 = **24**

 6 x 9 = **54**

 6 x 8 = **48**

 6 x 5 = **30**

7s Answer Key

 7 x 5 = **35**

 7 x 3 = **21**

 7 x 9 = **63**

 7 x 4 = **28**

 7 x 2 = **14**

 7 x 8 = **56**

 7 x 6 = **42**

 7 x 7 = **49**

8s Answer Key

 8 x 8 = **64**

 8 x 6 = **48**

 8 x 4 = **32**

 8 x 3 = **24**

 8 x 9 = **72**

 8 x 2 = **16**

 8 x 5 = **40**

 8 x 7 = **56**

9s Answer Key

 9 x 2 = **18**

 9 x 9 = **81**

 9 x 7 = **63**

 9 x 3 = **27**

 9 x 5 = **45**

 9 x 8 = **72**

 9 x 4 = **36**

 9 x 6 = **54**

MINI MATH CHECKS

11s Answer Key

 11 x 10 =
_____110_____

 11 x 6 =
_____66_____

 11 x 3 =
_____33_____

 11 x 7 =
_____77_____

 11 x 2 =
_____22_____

 11 x 12 =
_____132_____

 11 x 9 =
_____99_____

 11 x 8 =
_____88_____

 11 x 5 =
_____55_____

 11 x 11 =
_____121_____

 11 x 4 =
_____44_____

12s Answer Key

12 x 10 =
_____120_____

12 x 6 =
_____72_____

12 x 3 =
_____36_____

12 x 7 =
_____84_____

12 x 2 =
_____24_____

 12 x 12 =
_____144_____

12 x 9 =
_____108_____

 12 x 8 =
_____96_____

 12 x 5 =
_____60_____

 12 x 11 =
_____132_____

 12 x 4 =
_____48_____

Mixed Math Facts Assessments

After students have demonstrated mastery on the individual math facts, it's important to test them on all math facts mixed together. Students cannot be considered fluent with math facts until they know them in any order and anywhere they see them, not just when listed together on a Daily Quick Quiz or Mini Math Check.

Written Assessment Methods

Included in this book are three Mixed Multiplication Facts Tests and answer keys:

- facts from 0 to 5
- facts from 0 to 9
- facts from 0 to 12, with an emphasis on the more difficult math facts

Use the version of the mixed facts test that works best for your class.

LAURA'S Tips

Younger children may feel overwhelmed to be tested on all facts from 0 to 12 or even 0 to 9 together. I recommend starting with the math facts test of 0 to 5 for this group.

This book includes one Mixed Division Facts Test and its answer key. After students master the multiplication facts, most of them easily learn the division facts from 1 to 9, which are the facts needed for long division.

You can download an additional version of each of the Mixed Facts Tests from the online resources page, **www.lauracandler.com/mmf.** ONLINE

The first time you use any of these tests, you may want to eliminate the time limits and simply assign them for homework or independent practice. Later, you can add time limits to help students increase their fluency. The amount of time you set for students to complete each math fact test should be adjusted according to grade level and your own judgment. The following time limits are suggestions; feel free to adapt them as needed.

MULTIPLICATION TEST 0 TO 5 (40 PROBLEMS)

- Grade 3: four minutes
- Grade 4: three minutes
- Grade 5: two minutes

MULTIPLICATION TEST 0 TO 9 (40 PROBLEMS)

- Grade 3: five minutes
- Grade 4: four minutes
- Grade 5: three minutes

MULTIPLICATION TEST 0 TO 12 (50 PROBLEMS)

- Grade 3: seven minutes
- Grade 4: six minutes
- Grade 5: five minutes

DIVISION TEST 1 TO 9 (40 PROBLEMS)

- Grade 3: five minutes
- Grade 4: four minutes
- Grade 5: three minutes

Assessment Using Technology

Technology offers an alternative to written assessment methods. There are many online activities for practicing math facts, but I have found one in particular to be very helpful in assessing knowledge of math facts: Math Fact Café (**www.mathfactcafe.com**). You can have students use Math Fact Café to answer questions for a set of math facts, and then record their score in your Math Facts Record (**pages 133** or **134**). Technology may be especially helpful to assess the math facts knowledge of students with special needs.

Mixed Multiplication Facts Test
0 to 5

Name: _____ Date: _____

1. 7 x 1 _____ **15.** 5 x 1 _____ **29.** 9 x 1 _____

2. 6 x 1 _____ **16.** 0 x 5 _____ **30.** 4 x 7 _____

3. 7 x 2 _____ **17.** 6 x 4 _____ **31.** 4 x 9 _____

4. 4 x 2 _____ **18.** 9 x 3 _____ **32.** 3 x 4 _____

5. 3 x 3 _____ **19.** 4 x 5 _____ **33.** 7 x 3 _____

6. 0 x 7 _____ **20.** 2 x 6 _____ **34.** 4 x 4 _____

7. 5 x 8 _____ **21.** 2 x 8 _____ **35.** 6 x 5 _____

8. 2 x 3 _____ **22.** 2 x 7 _____ **36.** 3 x 1 _____

9. 5 x 2 _____ **23.** 3 x 8 _____ **37.** 0 x 8 _____

10. 1 x 4 _____ **24.** 3 x 0 _____ **38.** 7 x 5 _____

11. 2 x 2 _____ **25.** 8 x 4 _____ **39.** 2 x 1 _____

12. 8 x 1 _____ **26.** 9 x 5 _____ **40.** 5 x 3 _____

13. 2 x 9 _____ **27.** 3 x 6 _____

14. 0 x 6 _____ **28.** 5 x 5 _____

TIME _____ **SCORE** _____

MIXED MULTIPLICATION FACTS TEST
0 to 5
Answer Key

1. 7 x 1 __7__

2. 6 x 1 __6__

3. 7 x 2 __14__

4. 4 x 2 __8__

5. 3 x 3 __9__

6. 0 x 7 __0__

7. 5 x 8 __40__

8. 2 x 3 __6__

9. 5 x 2 __10__

10. 1 x 4 __4__

11. 2 x 2 __4__

12. 8 x 1 __8__

13. 2 x 9 __18__

14. 0 x 6 __0__

15. 5 x 1 __5__

16. 0 x 5 __0__

17. 6 x 4 __24__

18. 9 x 3 __27__

19. 4 x 5 __20__

20. 2 x 6 __12__

21. 2 x 8 __16__

22. 2 x 7 __14__

23. 3 x 8 __24__

24. 3 x 0 __0__

25. 8 x 4 __32__

26. 9 x 5 __45__

27. 3 x 6 __18__

28. 5 x 5 __25__

29. 9 x 1 __9__

30. 4 x 7 __28__

31. 4 x 9 __36__

32. 3 x 4 __12__

33. 7 x 3 __21__

34. 4 x 4 __16__

35. 6 x 5 __30__

36. 3 x 1 __3__

37. 0 x 8 __0__

38. 7 x 5 __35__

39. 2 x 1 __2__

40. 5 x 3 __15__

MIXED MULTIPLICATION FACTS TEST
0 to 9

Name:_____ Date:_____

1. 7 x 4 _____   15. 5 x 6 _____   29. 3 x 6 _____

2. 9 x 9 _____   16. 3 x 0 _____   30. 5 x 5 _____

3. 9 x 7 _____   17. 2 x 7 _____   31. 9 x 6 _____

4. 4 x 2 _____   18. 6 x 4 _____   32. 8 x 8 _____

5. 3 x 3 _____   19. 8 x 1 _____   33. 4 x 9 _____

6. 6 x 7 _____   20. 4 x 5 _____   34. 3 x 4 _____

7. 0 x 5 _____   21. 2 x 6 _____   35. 7 x 3 _____

8. 2 x 3 _____   22. 3 x 9 _____   36. 4 x 4 _____

9. 5 x 2 _____   23. 2 x 8 _____   37. 5 x 8 _____

10. 6 x 8 _____   24. 7 x 8 _____   38. 8 x 9 _____

11. 2 x 2 _____   25. 3 x 8 _____   39. 3 x 5 _____

12. 7 x 7 _____   26. 5 x 7 _____   40. 7 x 1 _____

13. 2 x 9 _____   27. 8 x 4 _____

14. 6 x 6 _____   28. 9 x 5 _____

TIME_____ SCORE_____

Laura Candler's Mastering Math Facts   •   Chapter 3   •   www.lauracandler.com

MIXED MULTIPLICATION FACTS TEST
0 to 9
Answer Key

1. 7 x 4 __28__

2. 9 x 9 __81__

3. 9 x 7 __63__

4. 4 x 2 __8__

5. 3 x 3 __9__

6. 6 x 7 __42__

7. 0 x 5 __0__

8. 2 x 3 __6__

9. 5 x 2 __10__

10. 6 x 8 __48__

11. 2 x 2 __4__

12. 7 x 7 __49__

13. 2 x 9 __18__

14. 6 x 6 __36__

15. 5 x 6 __30__

16. 3 x 0 __0__

17. 2 x 7 __14__

18. 6 x 4 __24__

19. 8 x 1 __8__

20. 4 x 5 __20__

21. 2 x 6 __12__

22. 3 x 9 __27__

23. 2 x 8 __16__

24. 7 x 8 __56__

25. 3 x 8 __24__

26. 5 x 7 __35__

27. 8 x 4 __32__

28. 9 x 5 __45__

29. 3 x 6 __18__

30. 5 x 5 __25__

31. 9 x 6 __54__

32. 8 x 8 __64__

33. 4 x 9 __36__

34. 3 x 4 __12__

35. 7 x 3 __21__

36. 4 x 4 __16__

37. 5 x 8 __40__

38. 8 x 9 __72__

39. 3 x 5 __15__

40. 7 x 1 __7__

MIXED MULTIPLICATION FACTS TEST
0 to 12

Name:_____ Date:_____

1. 7 x 4 _____

2. 9 x 9 _____

3. 7 x 9 _____

4. 11 x 3 _____

5. 4 x 12 _____

6. 1 x 9 _____

7. 6 x 7 _____

8. 5 x 8 _____

9. 12 x 3 _____

10. 5 x 12 _____

11. 6 x 8 _____

12. 12 x 2 _____

13. 7 x 7 _____

14. 11 x 8 _____

15. 12 x 9 _____

16. 6 x 6 _____

17. 11 x 12 _____

18. 5 x 6 _____

19. 3 x 5 _____

20. 12 x 7 _____

21. 6 x 0 _____

22. 8 x 9 _____

23. 11 x 2 _____

24. 12 x 12 _____

25. 12 x 6 _____

26. 3 x 9 _____

27. 12 x 8 _____

28. 7 x 8 _____

29. 3 x 8 _____

30. 11 x 6 _____

31. 5 x 7 _____

32. 8 x 4 _____

33. 9 x 5 _____

34. 3 x 6 _____

35. 5 x 5 _____

36. 9 x 6 _____

37. 8 x 8 _____

38. 4 x 9 _____

39. 11 x 10 _____

40. 7 x 3 _____

41. 4 x 4 _____

42. 11 x 11 _____

43. 10 x 12 _____

44. 10 x 10 _____

45. 9 x 2 _____

46. 11 x 9 _____

47. 7 x 6 _____

48. 8 x 10 _____

49. 4 x 6 _____

50. 3 x 3 _____

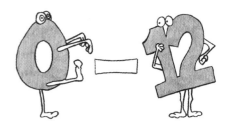

TIME _____ **SCORE** _____

Mixed Multiplication Facts Test
0 to 12
Answer Key

1. 7 x 4 __28__

2. 9 x 9 __81__

3. 7 x 9 __63__

4. 11 x 3 __33__

5. 4 x 12 __48__

6. 1 x 9 __9__

7. 6 x 7 __42__

8. 5 x 8 __40__

9. 12 x 3 __36__

10. 5 x 12 __60__

11. 6 x 8 __48__

12. 12 x 2 __24__

13. 7 x 7 __49__

14. 11 x 8 __88__

15. 12 x 9 __108__

16. 6 x 6 __36__

17. 11 x 12 __132__

18. 5 x 6 __30__

19. 3 x 5 __15__

20. 12 x 7 __84__

21. 6 x 0 __0__

22. 8 x 9 __72__

23. 11 x 2 __22__

24. 12 x 12 __144__

25. 12 x 6 __72__

26. 3 x 9 __27__

27. 12 x 8 __96__

28. 7 x 8 __56__

29. 3 x 8 __24__

30. 11 x 6 __66__

31. 5 x 7 __35__

32. 8 x 4 __32__

33. 9 x 5 __45__

34. 3 x 6 __18__

35. 5 x 5 __25__

36. 9 x 6 __54__

37. 8 x 8 __64__

38. 4 x 9 __36__

39. 11 x 10 __110__

40. 7 x 3 __21__

41. 4 x 4 __16__

42. 11 x 11 __121__

43. 10 x 12 __120__

44. 10 x 10 __100__

45. 9 x 2 __18__

46. 11 x 9 __99__

47. 7 x 6 __42__

48. 8 x 10 __80__

49. 4 x 6 __24__

50. 3 x 3 __9__

MIXED DIVISION FACTS TEST
1 to 9

Name:_____ Date:_____

1. $36 \div 6$ ____   **15.** $63 \div 9$ ____   **29.** $25 \div 5$ ____

2. $49 \div 7$ ____   **16.** $42 \div 7$ ____   **30.** $24 \div 4$ ____

3. $20 \div 5$ ____   **17.** $56 \div 8$ ____   **31.** $48 \div 8$ ____

4. $21 \div 3$ ____   **18.** $30 \div 5$ ____   **32.** $36 \div 9$ ____

5. $40 \div 5$ ____   **19.** $9 \div 1$ ____   **33.** $24 \div 8$ ____

6. $16 \div 8$ ____   **20.** $35 \div 7$ ____   **34.** $72 \div 9$ ____

7. $14 \div 7$ ____   **21.** $81 \div 9$ ____   **35.** $12 \div 4$ ____

8. $6 \div 1$ ____   **22.** $16 \div 4$ ____   **36.** $54 \div 6$ ____

9. $9 \div 3$ ____   **23.** $18 \div 6$ ____   **37.** $8 \div 2$ ____

10. $64 \div 8$ ____   **24.** $32 \div 4$ ____   **38.** $27 \div 9$ ____

11. $18 \div 9$ ____   **25.** $45 \div 5$ ____   **39.** $12 \div 6$ ____

12. $4 \div 2$ ____   **26.** $6 \div 3$ ____   **40.** $15 \div 5$ ____

13. $28 \div 4$ ____   **27.** $54 \div 9$ ____

14. $10 \div 2$ ____   **28.** $8 \div 1$ ____

TIME _____ **SCORE** _____

MIXED DIVISION FACTS TEST
1 to 9
Answer Key

1. 36 ÷ 6 ___6___   **15.** 63 ÷ 9 ___7___   **29.** 25 ÷ 5 ___5___

2. 49 ÷ 7 ___7___   **16.** 42 ÷ 7 ___6___   **30.** 24 ÷ 4 ___6___

3. 20 ÷ 5 ___4___   **17.** 56 ÷ 8 ___7___   **31.** 48 ÷ 8 ___6___

4. 21 ÷ 3 ___7___   **18.** 30 ÷ 5 ___6___   **32.** 36 ÷ 9 ___4___

5. 40 ÷ 5 ___8___   **19.** 9 ÷ 1 ___9___   **33.** 24 ÷ 8 ___3___

6. 16 ÷ 8 ___2___   **20.** 35 ÷ 7 ___5___   **34.** 72 ÷ 9 ___8___

7. 14 ÷ 7 ___2___   **21.** 81 ÷ 9 ___9___   **35.** 12 ÷ 4 ___3___

8. 6 ÷ 1 ___6___   **22.** 16 ÷ 4 ___4___   **36.** 54 ÷ 6 ___9___

9. 9 ÷ 3 ___3___   **23.** 18 ÷ 6 ___3___   **37.** 8 ÷ 2 ___4___

10. 64 ÷ 8 ___8___   **24.** 32 ÷ 4 ___8___   **38.** 27 ÷ 9 ___3___

11. 18 ÷ 9 ___2___   **25.** 45 ÷ 5 ___9___   **39.** 12 ÷ 6 ___2___

12. 4 ÷ 2 ___2___   **26.** 6 ÷ 3 ___2___   **40.** 15 ÷ 5 ___3___

13. 28 ÷ 4 ___7___   **27.** 54 ÷ 9 ___6___

14. 10 ÷ 2 ___5___   **28.** 8 ÷ 1 ___8___

CHAPTER 4

• • • • • • • • •

Math Facts
Practice Activities

Math Facts Practice Activities

• • • • • • • • •

After students understand multiplication and division conceptually, they need a variety of practice methods to increase speed and accuracy. The activities in this section are for use after students have mastered most math facts, to help them commit those facts to memory. They include group movement activities, flash card apps, engaging websites, times table challenges, and hands-on games that students play with a partner or a group. Your students will enjoy these fun activities and will look forward to your math fact practice sessions. Using the practice activities will help you meet Common Core Content Standard 3.OA.C.7: Fluently multiply and divide within 100.

- **FLASH CARDS (PAGE 174)**
- **MULTIPLICATION WITH MOVEMENT (PAGE 174)**
- **TIMES TABLE CHALLENGES (PAGE 177)**
- **TECHNOLOGY (PAGE 183)**
- **MATH GAMES (PAGE 184)**

Flash Cards

Flash cards might not be exciting, but they are definitely effective for studying and practicing math facts. Teach your students how to study with flash cards, both on their own and with a partner, using the *Flash Cards Study Method* steps (pages 175 - 176).

You can use store-bought flash cards or you can download a set from the Mastering Math Facts online resources page, **www.lauracandler.com/mmf**. Print them on card stock, write the answers on the backs, laminate them, and cut the cards apart. Store them in plastic zip bags. ✎ONLINE

LAURA'S Tips

You may want to print sets of the flash cards on different colored card stock. You can print an entire set of cards, 1 through 12, on one color and keep them together. Or you can print each math fact set on a different color. Color-coding the cards will help you return cards to their proper decks when you find them on the floor after class!

Multiplication with Movement

Many teachers have found that having students move while learning their math facts is an effective way to practice. Try these activities to get brains and bodies activated:

SKIP COUNTING ● How about skip counting, literally? Students form a circle and the teacher calls out a number such as "3." Students turn and skip clockwise as they chant the multiples of 3. Call out another number; students reverse direction and skip as they chant the multiples of the new number. Alternately, students use jump ropes and skip count as they skip rope.

BASKETBALL BOUNCING ● Have your students dribble a basketball in place as they skip count multiples of different numbers.

BEANBAG TOSS ● Students stand in a circle and the teacher calls out a single-digit number. Students toss the beanbag to each other while the entire group counts out the multiples of that number. The person who ends up holding the beanbag on the last multiple names a new number and the game continues.

FLASH CARDS STUDY METHOD

On Your Own:

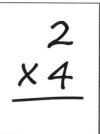

1 Shuffle the deck of flash cards and place them face up on your desk.

2 Silently read the first math fact and quietly whisper the answer to yourself.

3 Flip the card over to see if you are correct.

4 If you answered quickly and were correct, place the card in a "Facts Mastered" pile.

5 If it took a long time to answer or it was not correct, put the card back in the deck near the bottom.

6 Repeat until you have moved all the cards to the "Facts Mastered" pile.

FLASH CARDS STUDY METHOD

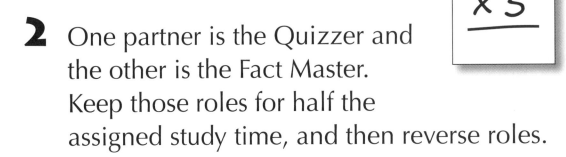

With a Partner:

1 Shuffle the deck of flash cards.

2 One partner is the Quizzer and the other is the Fact Master. Keep those roles for half the assigned study time, and then reverse roles.

3 The Quizzer holds up one card so the Fact Master can see the problem but not the answer.

4 The Fact Master quietly whispers the answer and the Quizzer checks by looking at the answer on the back of the card.

5 If the answer is correct, the card is put into a "Facts Mastered" pile. If the answer was not correct or was not given quickly, the Quizzer reveals the correct answer and the card goes back into the bottom of the deck.

6 Partners reverse roles halfway through the study time.

Times Table Challenges

Most teachers are familiar with times table charts, which are grids that show the factors in order along the left side and top, and have the products in the middle. Unfortunately, these charts are quite tedious for students to fill out and have limited value when it comes to math facts practice.

There are two variations of the times table chart that are much more interesting and challenging for students to complete. To make them self-checking, duplicate the answer keys that go with them and allow students to check their own work when finished.

MIXED-UP MULTIPLICATION ● The first time you use *Mixed-Up Multiplication* (page 178), display it in front of the class and ask your students if they see anything unusual about it. They are sure to notice that the factors along the top edge and left side are not in order. Demonstrate how to find the product of each row and column by having a volunteer come forward to complete the first row on the chart. After everyone grasps the concept, give each student his or her own copy of the chart to complete. If you would like to make your own mixed-up multiplication table, use the blank template (page 180).

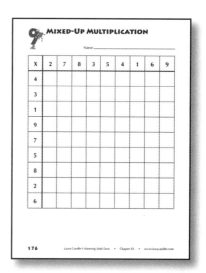

MYSTERY MULTIPLICATION ● *Mystery Multiplication* (page 181) is a more challenging version of *Mixed-Up Multiplication* because not only are the factors mixed up, students are required to find the unknown factor in a multiplication problem. This activity should be used after *Mixed-up Multiplication*; if you skip that activity, your students are likely to be extremely confused by *Mystery Multiplication*. On this chart, some of the factors and some of the products are missing. Students complete the chart by inferring the missing factor in each problem and filling it in on the table. Demonstrate the process by solving a few problems together. Point to the product of 8 in the first row and ask students to figure out what factor must be in the column above the 8. Since the 8 is in a row next to the factor of 2, students should think "What number times 2 equals 8?" The missing factor is 4, so they write the 4 in the box above the 8. Call on volunteers to demonstrate several more problems and then give each student his or her own copy to complete alone.

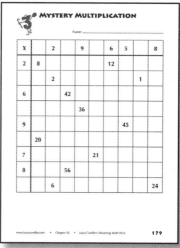

***Mystery Multiplication* aligns with Common Core Standards: for Content, 3.OA.A.4; for Practice, MP1, MP5, and MP7.**

MIXED-UP MULTIPLICATION

Name: _____

X	2	7	8	3	5	4	1	6	9
4									
3									
1									
9									
7									
5									
8									
2									
6									

Laura Candler's Mastering Math Facts • Chapter 4 • www.lauracandler.com

MIXED-UP MULTIPLICATION
Answer Key

X	2	7	8	3	5	4	1	6	9
4	8	28	32	12	20	16	4	24	36
3	6	21	24	9	15	12	3	18	27
1	2	7	8	3	5	4	1	6	9
9	18	63	72	27	45	36	9	54	81
7	14	49	56	21	35	28	7	42	63
5	10	35	40	15	25	20	5	30	45
8	16	56	64	24	40	32	8	48	72
2	4	14	16	6	10	8	2	12	18
6	12	42	48	18	30	24	6	36	54

MIXED-UP MULTIPLICATION

Name: _____

X									

MYSTERY MULTIPLICATION

Name: _____

X		2		9		6	5		8
2	8					12			
		2						1	
6			42						
				36					
9							45		
	20								
7					21				
8			56						
		6							24

MYSTERY MULTIPLICATION
Answer Key

X	4	2	7	9	3	6	5	1	8
2	8	4	14	18	6	12	10	2	16
1	4	2	7	9	3	6	5	1	8
6	24	12	42	54	18	36	30	6	48
4	16	8	28	36	12	24	20	4	32
9	36	18	63	81	27	54	45	9	72
5	20	10	35	45	15	30	25	5	40
7	28	14	49	63	21	42	35	7	56
8	32	16	56	72	24	48	40	8	64
3	12	6	21	27	9	18	15	3	24

Technology

Technology really shines when it comes to practicing math facts. Whether students use math practice websites, computer programs, or iPad apps, the format is generally more engaging and interesting than paper and pencil work. Technology is particularly effective for studying math facts because most activities provide immediate feedback about the accuracy of their student responses. Here are a few technology applications that I recommend. Some websites also have free downloadable programs or applications for mobile devices.

XTRAMATH (**www.xtramath.org**) ● This free website is highly recommended by many teachers who use it regularly with their students. Teachers can set up a class and enter each student's name individually. Students log on with a password and practice 10 or 15 minutes a day at home or at school. Both teachers and parents receive weekly progress reports.

ARCADEMIC SKILL BUILDERS (**www.arcademicskillbuilders.com**) ● This free website has loads of math facts practice games. Students can play against each other, as well as against the computer.

TIMEZ ATTACK (**www.timezattack.com**) ● Timez Attack is an exciting collection of video games for studying math facts. The graphics are terrific, and your students will love navigating a selected character through the different video-game environments to solve the problems.

MATH FACT CAFE (**www.mathfactcafe.com**) ● If you have access to a computer lab, you can use Math Fact Café for a speed drill race. Ask all your students to display the start page for a particular set of cards. After you say "Go!" they solve the problems as quickly as possible. If they miss one, they can start over. The first person to get 100% correct wins that round.

FLASH CARD WEBSITES AND APPS ● Flash card practice can be more engaging and effective using technology. The free flash card websites Study Blue (**www.studyblue.com**) and Brain Flips (**www.brainflips.com**), and the mobile app Math Cards! all work well.

Math Games

Math games are both fun and effective for reviewing basic math facts. These games rely more on skill than on luck, so students who know the math facts are rewarded by outperforming those who don't. These games can be played in cooperative learning teams or in math centers. You can even send the games home with students so they can play with a parent or sibling!

Many of the games included in this section deal specifically with multiplication. That's because most students have no problem learning the division facts after they master the times tables. Each game comes with complete student directions. Read the game directions and assemble a packet of materials prior to introducing it to your class. You may want to model the game for the entire class before placing it in a center or using it with cooperative learning teams. You can use decks of playing cards for these games, or print the cards with numbers from 1 to 12 in this book (page 185). For each set of 48 cards, print out four copies of page 185, laminate them, cut the cards apart, and store them in a plastic zip bag.

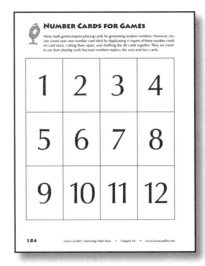

The key to using math games effectively is developing clear and specific classroom management procedures. Students need to know when they can play the games, where to go to play them, how to choose a partner, and a host of other procedures. If you are not familiar with using math games in your classroom, download the Tips for Teaching with Math Games in the online resources page, **www.lauracandler.com/mmf**. ONLINE

Math games in this chapter:

- **MATH FACT SHOWDOWN**
- **MULTIPLICATION 500**
- **MULTIPLICATION WAR**
- **IN THE DOG HOUSE**
- **TIC-TAC-TOE PRODUCTS**
- **MULTIPLICATION SHAKE-UP**

Number Cards for Games

Many math games require playing cards for generating random numbers. However, you can create your own number card deck by duplicating 4 copies of these number cards on card stock, cutting them apart, and shuffling the 48 cards together. They are easier to use than playing cards because numbers replace the aces and face cards.

1	2	3	4
5	6	7	8
9	10	11	12

MATH FACT SHOWDOWN

NUMBER OF PLAYERS: 2 to 4

Object of the game: Non-competitive game to practice math facts

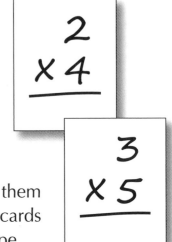

MATERIALS:

- Deck of flash cards
- Individual dry-erase boards and markers
- Calculator or math fact chart for checking answers

DIRECTIONS:

1 Before starting, shuffle the flash cards and place them face down in the middle of the playing area. If the flash cards have the answers on the back, place them in an envelope.

2 Decide who will be the first Leader.

3 The Leader turns over the top card or slides it from the envelope and reads the math fact aloud to the group. If the flash cards have answers on the back, the Leader reads the math fact aloud and leaves the card on the top of the pile during this turn.

4 Everyone on the team, including the Leader, writes the answer. No talking!

5 As soon as each player finishes writing the answer, he/she turns the dry-erase board face down.

6 When all the boards are face down, the Leader says, "Showdown!"

7 Everyone flips their boards over and shares their answers.

8 Celebrate correct answers and help anyone whose answer was not correct. You may have to use a calculator or consult a math fact chart to make sure you are correct.

9 The role of Leader rotates to the left and the game continues with the new Leader reading the top flash card.

10 When you are finished with the game, return the cards to the proper storage location.

MULTIPLICATION 500

NUMBER OF PLAYERS: 2 to 4

Object of the game: Win the race by being the first person to write the math facts correctly

MATERIALS:

- Deck of playing cards with aces and face cards removed, or 4 sets of number cards (2 - 9)

- Numbered race car cards (color each car a different color before playing the game for the first time)

- Individual dry-erase boards and markers or paper and pencil

- Calculator or multiplication chart for checking answers

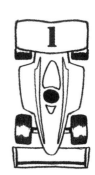

DIRECTIONS:

1 Before starting, shuffle the playing cards and place the stack face down in the middle of the playing area. Spread the race car cards face up next to the playing cards.

2 Decide who will be the first Leader.

3 The Leader turns over the top card and announces the number.

4 As quickly as possible, everyone writes out the multiplication facts for that number from 1 to 9 in order. See example.

7 x 1 = 7
7 x 2 = 14
7 x 3 = 21
7 x 4 = 28
7 x 5 = 35
7 x 6 = 42
7 x 7 = 49
7 x 8 = 56
7 x 9 = 63

5 The first person to finish turns his or her dry-erase board face down and takes race car #1. The next finisher takes race car #2, and so on, until everyone finishes.

6 Everyone checks their answers using a times table chart or calculator. The person who finished first with all answers correct scores a point. If the player who finished first had errors, the next player to finish correctly wins the round.

7 Place all race cars cards face up back in the center.

8 Repeat steps 3 through 7, rotating Leaders for each round.

9 The winner is the person with the most points at the end of the game.

Multiplication 500

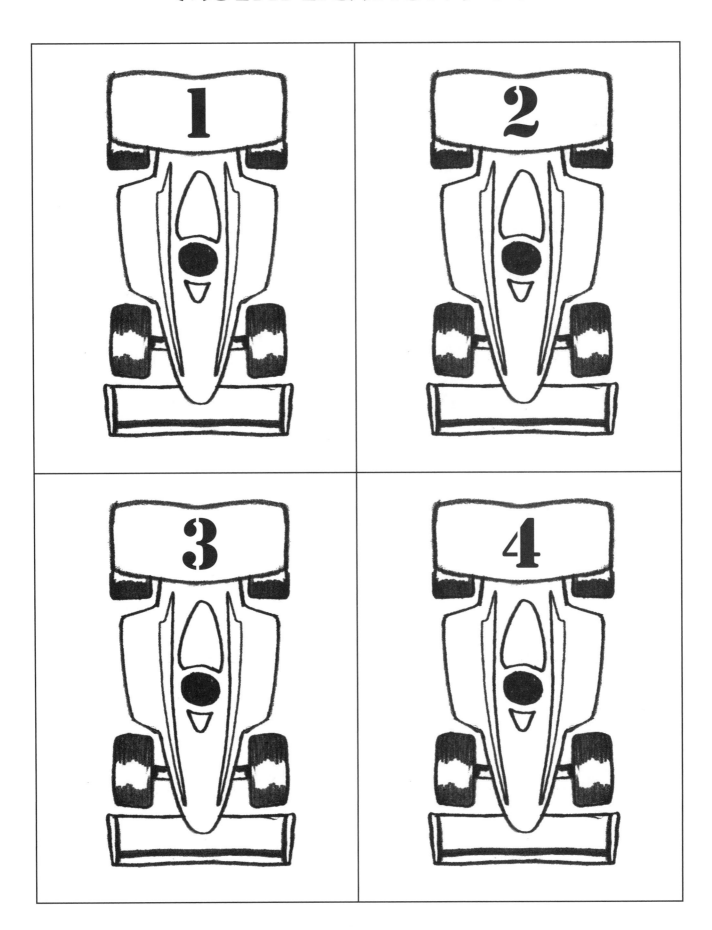

MULTIPLICATION WAR

NUMBER OF PLAYERS: 2

Object of the game: Capture the most cards by being the first to announce the product of the numbers on the cards

MATERIALS:

- Deck of playing cards with aces and face cards removed, or 4 sets of number cards (2 - 9)

DIRECTIONS:

1 Shuffle the deck of cards and divide them evenly between the two players. Do not look at the cards!

2 Players face each other. Both at the same time, players take the top card from his/her deck and place it face up between the players.

3 Immediately multiply the two numbers and call out the answer. Whoever says the product first wins both cards and places them at the bottom of his/her deck.

4 If both players say the product at the exact same time, they declare WAR! Together, both players chant the words, "I Declare War," and each places three cards on the table. Put the first two cards face down and the third card face up. Quickly multiply the two cards that are face up and say the product. Whoever says the product first wins all the cards in the war!

5 Repeat steps 1 - 4 until one person wins all the cards or the time is up. The person with the most cards is the winner.

IN THE DOG HOUSE

NUMBER OF PLAYERS: 2

Object of the Game: Win the most dog bones by recalling multiplication facts quickly and accurately

MATERIALS:

- In the Dog House game board

- 20 - 30 "dog bones" (paper clips, dried beans, etc.)

- Deck of playing cards with aces and face cards removed, or 4 sets of number cards (2 - 9)

DIRECTIONS:

1 Players face each other with the game board between them. Shuffle the cards. Place the deck face down in the Dog Pen in the middle of the board. Pile the dog bones (paper clips, dried beans, or other small items) on the dog bone pile.

2 Each player draws five cards from the Dog Pen. Players may look at their own cards but should not show them to their opponent.

3 To begin play, both players choose two of their own cards to place face down in their own Dog House. After both players have placed their cards, they turn them face up and multiply the numbers on their own two cards. Each player announces his or her product aloud. The player with the greatest product wins and takes a bone from the pile. If a player does not state the product correctly, the other player automatically wins the round.

4 Players remove both of their cards from the game board and set them aside to create a discard pile. Each player draws two new cards from the Dog House so that they each have five cards.

5 Repeat steps 3 and 4 until time is up. If all the cards in the deck are used, shuffle the discard pile and place it back in the Dog Pen.

6 The winner is the player with the most dog bones.

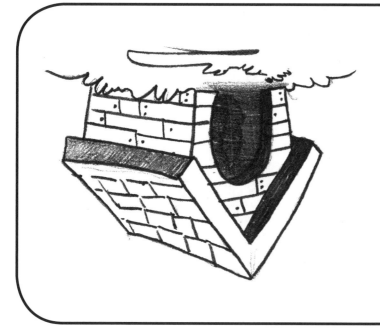

DOG HOUSE

DOG PEN

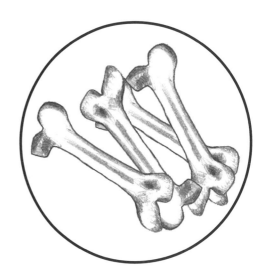

DOG HOUSE

TIC-TAC-TOE PRODUCTS
TWO PLAYERS

Object of the Game: Be the first player to cover 3 squares in a row horizontally, diagonally, or vertically

MATERIALS:

- Tic Tac Toe game board

- Game markers in 2 colors (checkers, bingo chips, colored cubes, etc.)

- Two large paper clips

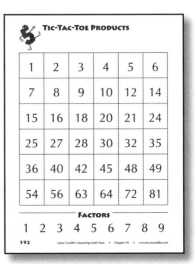

DIRECTIONS:

1 Both players begin by tossing a game marker onto the game board. The person whose marker lands on the highest number chooses a color and becomes Player 1.

2 To play, Player 1 slides the 2 paper clips onto the bottom edge of the game board so that they point to different factors (1 - 9).

3 Player 1 multiplies the two numbers and places a game marker onto the product on the game board.

4 Player 2 slides one paper clip to a different number and leaves one paper clip in place. Player 2 covers the new product with a different color marker.

5 Players take turns moving one paper clip and covering the product of the numbers.

6 The winner is the first player to cover 3 squares in a row horizontally, vertically, or diagonally.

Tic-Tac-Toe Products

1	2	3	4	5	6
7	8	9	10	12	14
15	16	18	20	21	24
25	27	28	30	32	35
36	40	42	45	48	49
54	56	63	64	72	81

Factors

1 2 3 4 5 6 7 8 9

MULTIPLICATION SHAKE-UP

TWO PLAYERS

Object of the Game: Be the first person to earn 500 points by finding the correct products

MATERIALS:

- Two small objects such as dried beans or paper clips

- Egg carton randomly numbered 1 to 9 or 1 to 12

- Scoring chart or paper for keeping score

- Calculator or multiplication chart

- Pencil

DIRECTIONS:

1 Decide who will be Player 1 and who will be Player 2. Make a T-chart for keeping score and write the two names at the top of the chart.

2 Player 1 begins by placing the 2 objects into the egg carton and closing the top. He or she shakes the carton gently and opens it up.

3 Player 1 calls out the two numbers marked by the objects and mentally multiplies them to find their product.

4 If Player 1 correctly names the product, that number is recorded on the chart as his or her points for that round. Use the calculator or multiplication chart to check the answer. If the product is not correct, he or she records a 0 for points earned.

5 Player 2 repeats steps 2 - 5 and records the points earned.

6 Continue taking turns shaking the carton and multiplying the numbers. After each round, use the calculator to add and record the total points. The winner is the first player to earn 500 or more total points.

7 The loser of the first game becomes Player 1 for the next game.

SCORING CHART

ABOUT THE AUTHOR

Laura Candler is a teacher with 30 years of classroom experience in grades 4 through 6. She has a Master's Degree in Elementary Education, National Board Certification as a Middle Childhood Generalist, and was a Milken Family Foundation Award winner in 2000.

Laura is the author of books and materials that help teachers implement new teaching strategies. Her work bridges the gap between educational theory and practice. Through her materials and her dynamic, interactive workshops, she gives teachers the tools they need to implement teaching strategies immediately.

Laura's materials are "field tested, teacher approved." They have been used by thousands of real teachers in real classrooms all over the world. Laura modifies and adapts her programs based on the experience of those teachers.

For more information and resources, go to Laura Candler's Teaching Resources website at www.lauracandler.com.

ABOUT COMPASS PUBLISHING AND BRIGANTINE MEDIA

Compass Publishing is the educational book imprint of Brigantine Media. Materials created by real education practitioners are the hallmark of Compass books. For more information, please contact:

Neil Raphel
Brigantine Media | 211 North Avenue | Saint Johnsbury, Vermont | 05819
Phone: 802-751-8802
E-mail: neil@compasspublishing.org | Website: www.compasspublishing.org

Laura Candler's
POWER READING WORKSHOP
A STEP-BY-STEP GUIDE

Reading opens up the world for your students.

Laura Candler's Power Reading Workshop: A Step-by-Step Guide will help you teach your students to love reading. Designed by award-winning teacher Laura Candler, this book walks you through the first ten days to implement a basic Reading Workshop with your students. Then Laura shows you how to add twelve proven "Power Reading Tools" to the program to make your Reading Workshop the most effective reading instruction you will ever use. Students and teachers alike love the simplicity, fun, and excitement of Laura's Power Reading Workshop program.

Everything you need, including reproducible worksheets, charts, and forms, are included to help you implement a Power Reading Workshop in your classroom today. Teachers around the world have used Laura's straightforward approach with fantastic results—that's what makes her program "field tested and teacher approved."

To purchase the book or digital download, or for more information, go to:
www.powerreadingworkshop.com

• • • • •

COMPASS
A DIVISION OF BRIGANTINE MEDIA